An ASIC Low Power Primer

T0143067

Rakesh Chadha • J. Bhasker

An ASIC Low Power Primer

Analysis, Techniques and Specification

 Springer

Rakesh Chadha
eSilicon Corporation
New Providence, NJ, USA

J. Bhasker
eSilicon Corporation
Allentown, PA, USA

ISBN 978-1-4899-9150-8   ISBN 978-1-4614-4271-4 (eBook)
DOI 10.1007/978-1-4614-4271-4
Springer New York Heidelberg Dordrecht London

Printed on acid-free paper

Springer is part of Springer Science+Business Media (www.springer.com)

Preface

How many times have you been ready to take a picture or a video when the battery in your device runs out? Invariably, it frequently happens to many of us. Even though the problem is not the lack of power in the battery, or conversely, that the unit has consumed too much power, guess who gets the brunt of the anger? In such moments, we always wish the camera or the video recorder did not consume so much power. However, it could be that, even in standby mode, the device was consuming a lot of power without our knowledge.

Most of us are now aware of the importance of power. From the giant server farms that consume large amounts of power to the smallest portable units such as a pacemaker that needs to last for a long time, the power requirement is a critical item of interest. For the giant server farms, they want to go "green," consume less power so that the operating costs as well as the impact on the environment are minimized. For the smallest portable units such as a pacemaker, you want the unit to last forever. The key to achieving all of these is to understand and analyze where the power is being consumed, have a way to measure the power, and finally adopt techniques that can reduce the amount of power consumed by the device or system.

In this book, we focus primarily on the CMOS digital ASIC devices. The book explores the three paradigms, how to analyze or measure power, how to specify the power intent for a device, and what techniques can be used to minimize the power consumption.

One of the challenges in measuring the power of an ASIC device is to figure out the conditions for worst-case power consumption. Is the power larger in cold conditions or in hot conditions? Is it when you press button A and button B together or is it when you press button A and button C together? Is the power worse while browsing the Internet, or is it worse when playing a video game? Is the power in standby mode also a large value? These questions indicate that there is a concept of the worst power scenario. It is likely that a user may never operate under such a scenario. So do you really need to design the device to handle such conditions? Or should you target the power to be minimized for typical applications? An ASIC system designer has his or her work cut out, as these are not easy questions to answer. For example, an MP3 player was not power-optimized for playing video songs. If it plays only audio songs, the battery lasts for 4 days. If it plays video songs, the battery runs out within 6 h.

This book is targeted towards professionals working on ASIC designs. A background in logic design is required. It has been written in an easy-to-read style where there are almost no dependencies amongst the chapters. You can jump into the chapter of interest straightaway. Initial chapters are focused on explaining how to measure power. Later chapters describe the implementation strategies to reduce power, and finally, we describe the languages that can be used to describe the power intent.

Acknowledgments

We would like to express our deep gratitude to eSilicon Corporation for providing us the opportunity to write this book.

We also would like to thank and acknowledge the valuable feedback provided by Marc Galceran-Oms, Pete Jarvis, Luke Lang, Carlos Macian, Ferran Martorell, Satya Pullela, Prasan Shanbag, Hormoz Yaghutiel, and Per Zander. Their feedback has been invaluable.

Last but not least, we would like to thank our families for their patience during the development of this book.

New Providence, NJ, USA   Rakesh Chadha
Allentown, PA, USA   J. Bhasker

Contents

Chapter 1
Introduction

The power consumption of a semiconductor device is a critical design parameter that governs the design, implementation and usage within the system. Along with the target speed, the power consumption is a key performance metric of the device. The power consumption dictates the power as well as the thermal requirements (or heat dissipation) of the package.

Historically, the semiconductor device design has focused on getting the target performance within the specified silicon area. This implies that the semiconductor device designers focused upon squeezing as many transistors as possible within the smallest silicon area while meeting the target performance specification.

1.1   What Is Power?

Any electronic device draws current when it is powered *on*. The amount of current drawn depends upon whether the device is in normal functional mode or in standby mode. For example, a mobile phone draws larger current from its battery when the user is making a call (that is, the device is in functional mode) than when the user is not using the phone (that is, the device is in standby mode). In either scenario, power is being consumed and the battery is getting drained. Similarly, a laptop computer draws current from its battery or, when it is plugged in, draws power from the main supply.

In each of the above scenarios, the semiconductor devices are consuming power from its power supply.

1.2   Why Is Power Important?

It is important to understand the power requirements for semiconductor devices. It is well known that power drawn is critical for battery-operated devices as it controls the battery life (time before a recharge is required). There are other equally important

R. Chadha and J. Bhasker, *An ASIC Low Power Primer: Analysis,*
Techniques and Specification, DOI 10.1007/978-1-4614-4271-4_1,
© Springer Science+Business Media New York 2013

considerations regarding power consumption for both battery-operated devices and for devices powered from the main supply:

1. The power consumed by a semiconductor device is dissipated within the device resulting in an increase in temperature inside the device.
2. The device internal temperature should be constrained to an acceptable range for proper operation and reliability. This requires that the heat generated has to be drained away imposing requirements on the package and potentially requiring the use of a heat sink and/or a fan.
3. The power availability within a system can be limited. For example, only limited amount of power is available in a system powered by Ethernet (PoE[1]). Other examples are systems with the backplane power budget limit or systems where cost of power supply unit limits the available power.

In view of the above, it is critical to understand the power consumption so that a designer can evaluate various system trade-offs related to the device and the operating environment. Some examples of trade-offs are provided below:

1. *Battery life vs battery size*: This is dependent on the total energy consumption until requiring a recharge. Batteries have to be selected which can supply the required power while providing adequate battery life. In general, higher power dissipation would require bigger batteries which can increase the cost and add to the weight (critical for handheld applications).
2. *Peak power dissipation*: Manage the heat dissipation from the peak power. This involves selecting, amongst others, the optimal package, the heat sink and the fan.
3. *Performance/power trade-offs*: Evaluate the performance power trade-offs for the system. This involves determining the target process technology and the performance of the device. Since the power dissipation and the achievable performance are dependent on the process technology, the designer needs to evaluate multiple dimensions while evaluating these trade-offs.
4. *Achieving the target performance*: The application must provide a response time that is acceptable to the user.

1.3 Why Is Power Increasing?

The computational power of computers has been growing as the designers have managed to squeeze increasingly larger number of transistors into a silicon device and are also switching those transistors at increasingly faster speeds. Based upon the Moore's law [MOO65], the number of transistors in an integrated circuit approximately doubles every 2 years. Every new generation of semiconductor process technology doubles the transistor density which implies that twice as many transistors get packed in

---

[1] Power Over Ethernet.

Table 1.1 Core power supply and logic density for different process technology nodes

Technology	Core power supply (V)	Gate density (per mm²)
90 nm	1.0	354 K
65 nm	1.0	694 K
40 nm	0.9	1,750 K
28 nm	0.85	3,387 K

the same silicon area. With a new process technology being introduced every 2 or 3 years, the number of devices on a chip keeps increasing exponentially.

While the voltage supply has reduced from the early generations, it has not reduced fast enough and we seem to have hit a limit of ~0.8 V for the core supply. Since the power supply voltage has not reduced, the power being dissipated in a device keeps increasing. Similarly, the interconnect trace length has reduced at a smaller rate than the increase in gate density. While newer low-K materials and shorter trace lengths have reduced the capacitance load of a transistor, the overall power of the device keeps increasing since the designers push the operating frequency of the device higher and higher. An example of the process technology and the power supply is given in Table 1.1.

Another factor leading to the increase in power of a device is the increase in leakage. This is the component of power dissipation in a device which is present even when the device is not doing any computation. Whenever the device is powered *on,* the leakage power is being dissipated (in addition to any active power). As the technology shrinks, the leakage power has been increasing. This is because the leakage is related to the MOS transistor threshold voltage which is reducing with every new generation of process technology. Lower threshold voltage causes greater leakage which basically contributes to the power dissipated when the device is turned *on.*

In many of the high-end processors, too much power is required to keep all the transistors operating at full speed, implying that the device can potentially overheat if all the transistors are operated simultaneously. Some devices shut down a part of its functionality to stay within the power budget [ESM11]. The portion of a chip which is turned *off* is sometimes labelled as *dark silicon* much like a city suffering blackouts due to power shortage.

1.4 Where Is the Power Going?

The power dissipated in a device can broadly be classified as *functional* and *standby* power. The *standby* power is the power dissipated when the device is powered *on* but is not performing any computations. The *functional* power is the power dissipated by the device when it is performing the intended computations. The functional power can be considered to be comprised of the *static* power and the *dynamic* power. The *static* power is the power dissipation when there is no activity in the device and the *dynamic* power is the additional power due to any activity in the device. Since the goal is to keep the total power within the available budget, the total of the dynamic and static power determines the package and heat dissipation constraint.

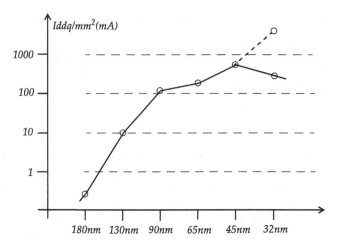

Fig. 1.1 Worst corner leakage per unit area for different process technology nodes

$$\text{Total power} = \text{Dynamic power} + \text{Static power}$$

For digital CMOS designs, the static power is the same as leakage power[2] while the dynamic power may also be referred to as active power.

For advanced technologies, the leakage power can be a major contributor to the total power especially at the worst-case leakage corner. See Fig. 1.1 for a representative comparison between a process node and *Iddq*; *Iddq* represents the leakage current at the worst case corner. As seen in this figure, the *Iddq* per unit area is increasing as the process geometry shrinks. One exception is the 32 nm process node where the use of the metal gate reverses the trend line of *Iddq* increase from previous generations.

The use of battery-powered devices is leading to various design and implementation techniques to reduce power. Even the non-battery powered devices have several reasons to reduce power; some of these are described in Sect. 1.2, and another reason can be that they stay *green*.

1.5 How Much Is Low?

Various applications in widely different domains have requirements for low power, even though the amount of power may vary widely from one application to another. Two examples with very different power requirements are described below.

Consider a parallel processing system utilizing ~24 modules operating in parallel where each module contains 12 printed circuit boards and each board contains 16

---

[2] The leakage and dynamic power for CMOS designs is described in detail in Chap. 2.

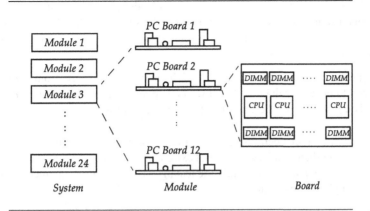

Fig. 1.2 An example parallel processing system

processing devices plus 32 DIMMs[3] mounted on it. The power utilized by such a system (illustrated in Fig. 1.2) would be ~60 kW. Such a system is obviously powered from the main power supply and the key constraint is to be able to keep it cool, that is, to ensure that it does not overheat due to the ~60 kW power being dissipated in the system. Reducing the power requirement without degrading the system processing throughput can result in a large savings in cooling and power cost.

Now let us consider another example of a mobile telephone. The mobile telephone dissipates power all the time as it is always *on*. The power consumption is ~100 mW when the user is making or receiving phone calls and the consumption reduces to less than 1 mW when it is not in use. Keeping the power consumption low is important since the user should not need to recharge the mobile phone battery frequently (enabling the mobile phone manufacturers to claim that the phone can be used for several days without requiring a recharge). Lower power consumption also helps to reduce the battery requirement; note that a lightweight and a small battery enables the mobile phone to fit into a sleek package.

Thus, whether the system uses milliwatts of power or kilowatts of power, it is always critical to have a low power system. In each category, the low power can mean different savings—whether reducing a few kilowatts in a ~60 kW system power or reducing a few milliwatts in a tablet computer whose power budget is 100 mW.

1.6 Why Measure?

Like in any other system, the key to achieving low power is to first determine the amount of power consumed in various usage modes. The designer of the mobile phone system must analyze the power consumption in these modes. The design

---

[3] Dual in-line memory module (typically DRAMs).

team of a tablet device would analyze not only the power consumption in these modes but also how the power consumption varies due to the variations in operating environment or due to the variation in semiconductor devices used in the system. For example:

1. The power consumption of a tablet which happens to use parts from a *fast wafer lot* would be higher than the tablet which happens to use parts from a *slow wafer lot*. The *fast* and *slow* here refer to the statistical variations in the manufacturing process.
2. Similarly, a tablet operating in a cold and freezing temperature would require lower power than a tablet operating at extremely hot temperatures. This is because the power dissipation (especially the leakage or static power) can increase many-fold at higher temperatures.

A power analysis of the system enables the designer to choose the right architecture and then select the appropriate process technology for manufacturing the device(s) used in the system. An accurate power estimate also helps the system designer in verifying the thermal constraints so that the system does not overheat (that is, the device temperature stays within acceptable limits).

1.7   Impact on Design Complexity

The design complexity is growing and keeping pace with the increasing integration offered with every process generation. To manage the power constraints, a typical power-hungry design is composed of multiple power domains. See Fig. 1.3 for an example. There are four different power domains in this device. The *CPU* domain is always-on and runs at 1.1 V, the *SRAM* domain is always-on but runs at 0.9 V, while the coprocessor and the memory controller domains operate at 0.9 V but can be shut down.

Timing closure after physical design is tougher because of many power domains. Multiple iterations are likely needed between sign-off corner analyses and for optimization. There is also the additional challenge to determine the size and the number of power switches to be used, because this has to be based on a good post-layout power estimate. Few trials to determine the optimal use and layout of the power switches may be required. Obtaining good quality of results in optimization becomes harder and may involve multiple iterations. The complete design flow takes longer due to the additional task of handling switched domains and multiple voltage domains.

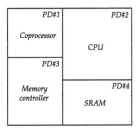

Fig. 1.3 A design with four different power domains

1.8   Outline of the Book

This book provides an in-depth review of the techniques utilized in the design of low power digital semiconductor devices. Since analyzing the system power consumption is a prerequisite to low power design, Chaps. 2–4 are focused on the power analysis of a digital CMOS device. Chapter 2 describes the modeling of power dissipation in core standard cell logic. The techniques for modeling the IO buffers and SRAM macros are described in Chap. 3. Based upon the modeling of individual components described in Chaps. 2 and 3, the procedures for computing the power dissipation in a design are described in Chap. 4.

Chapters 5–7 are focused on techniques that can be employed in a low power design. Chapter 5 details on how a designer can provide the high level intent for managing the power consumption for a design. The design intent is in terms of whether some portions of the design can be switched off and/or operated at reduced power supply to trade-off power with performance. Since most of the power savings in a design are dependent on choosing the optimal architecture, Chap. 6 describes the techniques that can be adopted at the architectural design phase of a system. Based upon the chosen architecture, the low power implementation techniques are described in Chap. 7.

Chapters 8 and 9 describe two alternate standards for capturing the power directives at various phases of the design. Chapter 8 describes the Unified Power Format (UPF) which is an IEEE standard. Chapter 9 describes the Common Power Format (CPF) which is an alternate power specification language.

The two appendices at the end provide the details of the Switching Activity Interchange Format (SAIF) and the UPF. Appendix A contains detailed syntax of SAIF and Appendix B describes the UPF syntax in detail.

Chapter 2
Modeling of Power in Core Logic

This chapter describes various aspects of power dissipation in core digital logic in a CMOS design. The power dissipation in an ASIC is comprised of power in the digital core logic, memories, analog macros, and other IO interfaces. The power dissipation in digital logic and memory macros can be due to switching activity, called *dynamic power*, and due to a contribution called *leakage power*. This chapter describes the modeling of these contributions for core logic—specifically the factors affecting the power calculation from the standard cell logic in the design.

2.1 Power Dissipation in Digital Designs

To understand the power dissipation in digital designs, it is instructive to consider the example of the capacitor charging and discharging through an ideal switch in series with a resistor.

2.1.1 Example Using Ideal Switch

Consider the charging and discharging of a capacitor. Figure 2.1 shows the capacitor *C* connected to a power supply *Vdd* with resistance *Rpu* and a switch *SW1*.

Assuming the capacitor is initially uncharged, the closing of switch *SW1* causes the capacitor to start charging towards the final voltage *Vdd*. The power supply provides the current until the capacitor *C* is fully charged to *Vdd*. During the charging of the capacitor, the total energy provided by the power supply is given by:

$$E_{total} = C * Vdd^2$$

One-half of the above energy is dissipated in the charging resistor and the other half is transferred to the capacitor. The energy transferred to the capacitor, E_{cap}, is given by:

R. Chadha and J. Bhasker, *An ASIC Low Power Primer: Analysis,*
Techniques and Specification, DOI 10.1007/978-1-4614-4271-4_2,
© Springer Science+Business Media New York 2013

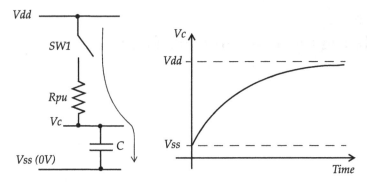

Fig. 2.1 Charging of a capacitor

$$E_{cap} = E_{diss} = C * Vdd^2 / 2$$

In above, E_{diss} refers to the total energy dissipated in the charging resistor. Now consider this capacitor C being connected to ground Vss through another resistor in series. Figure 2.2 shows the same capacitor C (now charged to Vdd) connected to ground Vss via the switch $SW0$ and resistor Rpd. The switch $SW1$ is considered open during the discharging of the capacitor.

When the switch $SW0$ is closed, the capacitor C discharges via resistor Rpd to Vss. When the capacitor is discharged, all the energy in the capacitor is dissipated through the resistor. Thus, the energy dissipated through the resistor is:

$$E_{diss} = C * Vdd^2 / 2$$

The example in Figs. 2.1 and 2.2 illustrates the charging and discharging of a capacitor C for various nets in a design for each cycle. Note that the total energy provided by the power supply during charging ($= C * Vdd^2$) is dissipated during the charging and discharging cycle. Also note that this energy depends only upon the capacitance value and power supply and does not depend upon the pull-up or the pull-down resistance.

2.1.2 CMOS Digital Logic

The illustration in the previous section of an idealized switch can be extended to CMOS digital designs. In general, the CMOS digital logic is comprised of combinational and sequential logic gates. In each logic gate structure, the switch to Vdd is realized by a PMOS pull-up structure and the switch to Vss is realized by a complementary NMOS pull-down structure. As an example, the equivalent pull-up and pull-down structures for a two-input *nand* gate are shown in Fig. 2.3. For more details, the reader is referred to a standard textbook describing CMOS logic [MUK86].

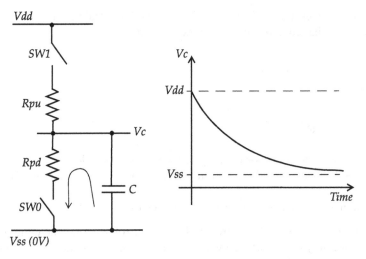

Fig. 2.2 Discharging of a capacitor

Fig. 2.3 CMOS two-input *nand* gate

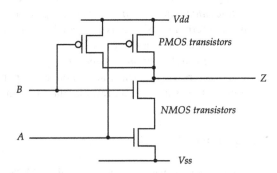

This section extends the concepts described in the previous section to power dissipation in CMOS digital designs. The complementary nature of the pull-up and pull-down implies that these structures can be treated as equivalent to the switches. However, there are two key differences:

(a) The turning *on* and turning *off* of the pull-down and pull-up structures is not instantaneous. Thus, during the transition, there is a small interval of time when both the pull-up and the pull-down structures are *on*.

(b) Even when the pull-up or the pull-down structures are *off*, there is a small conductance through the *off* structure that provides a current path between the power rails. The current flow through an *off* device contributes to leakage power which is the power dissipated even when the logic gate is not transitioning.

The equation for energy dissipated in the charging and discharging of a capacitor described in the previous section is also applicable to a CMOS gate and is referred to as *output charging power*. However, due to the non-idealities described above, there are two other components to the power dissipation:

1. Leakage power
2. Internal switching power

2.1.2.1   Leakage Power

Leakage or static power is power dissipated when the CMOS logic is not switching. As described above, this is due to the current flow through the *off* devices (PMOS or NMOS).

The non-zero leakage currents are mainly due to:

(a) *Subthreshold leakage*: Due to reduced threshold voltage, the source-drain channel is not fully turned *off* even when the gate of the NMOS device is connected to *Vss* (or when the gate of the PMOS device is connected to *Vdd*). Thus, any voltage difference between the *source* and the *drain* results in current flow through the drain-source channel of the MOS device which is *off*.
(b) *Gate-oxide tunneling current*: The charge can tunnel through the gate oxide and this is referred to as gate oxide tunneling current. This current can flow from a gate terminal into an *on* device or an *off* device.
(c) *Reverse biased junction leakage current*: There can be leakage between the diffusion layers and the substrate. Though the p–n junctions between the source/drain and the substrate are reverse-biased, a small current can still flow.

2.1.2.2   Internal Switching Power

This refers to the power dissipated when the CMOS logic gate is active, or switching. The major cause is the power dissipated during the brief interval of time when both the pull-up and the pull-down structures are *on*. This is also referred to as the *short circuit power*. Figure 2.4 shows a CMOS inverter gate when the signal at the input pin rises.

The waveforms represented in Fig. 2.4a, b are for the rising input to the CMOS inverter. Based upon the rise time of the input signal, the PMOS and NMOS devices are both *on* for a small interval of time (depicted as *Toverlap* in Fig. 2.4).[1] During the time both the PMOS and NMOS devices are *on*, the current flowing from the PMOS devices straight through to the NMOS devices does *not* contribute to the charging or discharging of the output capacitance. This current is called the *crowbar current* and may also be referred to as *overlap current*. The two sets of waveforms illustrate that the short circuit power dissipation will be larger in the case where input *rise* time is very slow (waveform in Fig. 2.4b).

---

[1] It is possible that the sum of PMOS and NMOS threshold voltages (*VtNMOS + VtPMOS*) exceeds the power supply voltage (*Vdd*). In such cases, both PMOS and NMOS can not be turned *on* together and there is no overlap current.

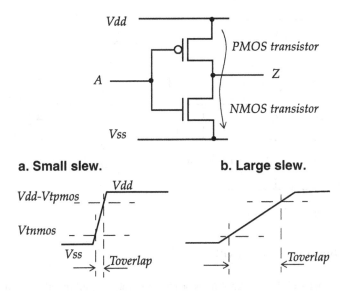

Fig. 2.4 CMOS inverter and rising waveforms at its input (**a**) small slew (**b**) large slew

It is not necessary for an input change in state to cause an output to switch; a small amount of power is still dissipated when an input to a CMOS logic gate switches and it does not result in output switching. This component of power is referred to as the *input pin power*.

The internal switching power does not include the power due to the switching of the output load capacitance which is considered separately.

2.1.2.3   Output Charging Power

The output charging power is the same as described using the example with an ideal switch in the previous section. In CMOS digital logic, the output charging power is the power dissipated due to the charging and discharging of the capacitive load at the output. The load capacitance at the output is due to the capacitance in the interconnect and the input pin capacitances of the fanouts. An example of the charging and discharging of the output is shown in Fig. 2.5. The CMOS logic gate *G1* is driving the inputs of three other gates, *G2*, *G3* and *G4*.

The activity in the CMOS digital design maps to nets in the design changing logic state from logic-0 to logic-1 or from logic-1 to logic-0. The switching of the CMOS logic charges and discharges its output to the logic-1 and logic-0 state respectively. For logic-0 to logic-1 output switching, the output node is charged through the CMOS pull-up devices and the interconnect resistance in the path (the interconnect resistance in the path is not shown in Fig. 2.5). Similarly, for logic-1 to logic-0 output switching, the output net is discharged through the interconnect resistance and the CMOS pull-down devices. This is, in principle, similar to the simple RC circuit illustrated through Figs. 2.1 and 2.2. The power dissipated does not depend upon the equivalent MOS resistance and depends only upon the power supply and the capacitor value.

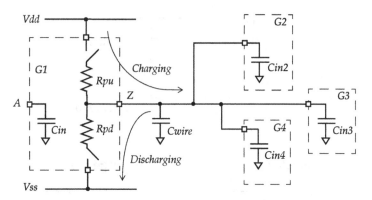

Fig. 2.5 Charging and discharging of output load

2.1.2.4   Multi-threshold CMOS

The finer geometry CMOS technologies provide a choice of multiple threshold PMOS/ NMOS transistors within the design. The lower threshold MOS transistors provide higher performance. The increased performance comes with a price of increase in leakage power. With the choice of multiple thresholds, the design can use lower threshold transistors (or cells) along the path where higher performance is critical and use higher threshold cells for the non-critical timing portions of the design. The use of multiple threshold transistors within a design is often referred to as *MTCMOS logic*.

2.1.2.5   Energy vs Power

Any introductory text on Physics can provide the basic background on energy, work and power. Energy is the ability of a system to perform certain work. It is measured in Joules (SI unit for energy). Power is the rate at which energy is transferred. Power is measured in watts (SI unit for power), which is the same as Joules per second.

With the broad overview of power dissipation in CMOS logic described in this section, the subsequent sections provide the details of the modeling of power in digital designs.

2.2   Dynamic or Active Power

The *internal switching power* and the *output charging power* components described in the previous section together comprise the *dynamic* or the *active power*. The discussion in the previous section used CMOS logic gates to illustrate the concept of internal switching power and output charging power. In a general scenario, a CMOS digital design is comprised of standard cells and other macros such as memory instances. The output charging power is directly obtained in terms of output load capacitance, and the internal switching power is obtained from the corresponding models included in the library description. The internal power information for standard cells and other macros is described using a library modeling standard such as Liberty [LIB]. While any standard modeling approach can be utilized to describe power, we have used the Liberty modeling for the examples within this book.

The output charging power is independent of the cell type and depends only upon the output capacitive load, frequency of switching and the power supply of the cell. The internal switching power includes any power dissipated within the cell due to charging and discharging of the internal node capacitances. The internal switching power depends upon the type of the cell and its model is included in the cell library. The specification of the internal switching power in the library is described next.

2.2.1 Active Power in Combinational Cells

For a combinational cell, an input pin transition can cause the output to switch which results in internal switching power. For example, an inverter cell consumes power whenever the input switches (has a rise or fall transition at the input).

Here is an example of the internal power specified on the output pin of a *nand* cell as described in a library.

```
pin (Z1) {
    .  .  .
    power_down_function :  "!VDD + VSS";
    related_power_pin : VDD;
    related_ground_pin : VSS;
    internal_power () {
        related_pin : "A";
        power (template_2x2) {
            index_1 ("0.05, 0.1"); /* Input transition */
            index_2 ("0.1, 0.25"); /* Output capacitance*/
            values ( /*            0.1            0.25 */ \
            /* 0.05 */            "0.045,         0.050", \
            /* 0.1 */             "0.055,         0.056");
        }
    }
}
```

The example above shows the internal power dissipation for a transition at the output pin *Z1*. The output transition is caused by a transition at the cell input pin *A*. The 2×2 table in the template is in terms of input transition at pin *A* and the output capacitance at pin *Z1*. Note that while the table is in terms of total output capacitance, the table values only correspond to the internal switching power and do not include the contribution due to the output capacitance. The values represent the internal energy dissipated in the cell for each switching transition (rise or fall). The energy units can be derived from other units in the library; typically voltage is in volts (V) and capacitance is in picofarads (pF), and this maps to energy in pico-Joules (pJ). The internal power in the library thus actually specifies the internal energy dissipated per transition.

Example Computation of Active Power: Consider an example with the *nand* cell whose power tables are as illustrated above. Assume that the input pin *A* has an input transition time of 0.075 ns and pin *Z1* has an output capacitance of 0.25 pF. The input pin *A* has a clock with frequency 100 MHz applied to it. The other input of the *nand* cell is assumed to be steady *high*. What is the total active (or dynamic) power for a *Vdd* of 1.0 V?

Based upon the internal power table above:

Energy dissipated per transition (contributing to internal switching power)
   = **0.053pJ per transition**

Internal switching power = 0.053 * 2 * 100 * 1E6
   = **10.6µW**

Output charging power $(CV_{dd}^2 f)$
   = 0.25*(1E-12)*(1E8) = **25µW**

Total active power (internal plus output)
   = **35.6µW**

The 0.053 pJ comes from the interpolation to the appropriate slew load point in the power table. The factor of 2 in the internal switching power accounts for both the rise and fall edges.

In addition to the power tables, the library excerpt for the internal power also illustrates the specification of the power pins, ground pins, and the power down function which specifies the condition when the cell can be powered off. These constructs allow for multiple power supplies in the design and for scenarios where different supplies may be powered down. The following illustration shows the power pin specifications for a cell.

```
cell (NAND2) {
  . . .
  pg_pin (VDD) {
    pg_type : primary_power;
    voltage_name : COREVDD1;
    . . .
  }
  pg_pin (VSS) {
    pg_type : primary_ground;
    voltage_name : COREGND1;
    . . .
  }
}
```

The power specification syntax allows for separate constructs for rise and fall power (referring to the output sense). The power specification can also be state-dependent. For example, the state-dependent power dissipation for an *xor* cell can be specified as dependent on the state of its two inputs.

In a multi-input combinational cell, an input pin can have a transition which does not cause the output to transition. An example is a transition at one of the inputs of an *and* (or *nand*) cell when another input pin is at logic-0. Even though the *and* (or *nand*) cell output does not transition, a small amount of internal power is dissipated in the cell due to the charging and discharging of internal node(s) within the cell. The internal power dissipated, by an input transition which does not result in an output transition, is specified separately within the input pin section of the cell description. Normally, the state-dependent *when*-condition qualifies the input power specification. This can be illustrated for a two-input *nand* cell as follows.

```
pin(A1) {
   direction : input;
   related_ground_pin : VSS;
   related_power_pin : VDD;
      . . .
   internal_power () {
   when :  "!A2&ZN";
   related_pg_pin : VDD;
   rise_power (passive_power_template_4x1_0) {
    index_1 ("0.002, 0.013,
               0.052, 0.238");
    values ( \
      "0.0242, 0.0213,
        0.0253, 0.0292" \
    );
   }
   fall_power (passive_power_template_4x1_0) {
    index_1 ("0.002, 0.013, 0.052,
          0.238");
    values ( \
      "0.0592, 0.0574,
        0.0597, 0.0587" \
    );
   }
  }
 }
```

A similar description is provided for the other input pin also.

Note that for combinational cells, the internal power dissipated in the cell when the output does not switch is relatively small in comparison to the scenario when the output of the cell also switches.

2.2.2   Active Power in Sequential Cells

As seen in the previous section, the switching power can be specified on an
input–output pin pair basis (for scenarios where the output pin switches) or as
part of the input pin description (for scenarios where the input transition does
not result in an output transition). However for a sequential cell such as a flip-
flop with complementary outputs Q and QN, the CLK->Q transition also results
in a CLK->QN transition. Thus, the library can specify the internal switching
power as a three-dimensional table, which is shown next. The three dimensions
in the example below are the input slew at CLK and the output capacitances at
Q and QN respectively.

```
pin (Q) {
   . . .
  internal_power() {
  related_pin : "CLK";
  equal_or_opposite_output : "QN";
  rise_power(energy_template_3x2x2) {
    index_1 ("0.02,    0.2, 1.0"); /* Clock trans*/
    index_2 ("0.005,   0.2"); /* Output   Q cap*/
    index_3 ("0.005,   0.2"); /* Output   QN cap*/
    values (/*0.005    0.2 */ /* 0.005    0.2 */ \
    /* 0.02*/"0.060,   0.070",   "0.061,   0.068", \
    /* 0.2 */"0.061,   0.071",   "0.063,   0.069", \
    /* 1.0 */"0.062,   0.080",   "0.068,   0.075");
  }
  fall_power(energy_template_3x2x2) {
    index_1 ("0.02,    0.2, 1.0");
    index_2 ("0.005,   0.2");
    index_3 ("0.005,   0.2");
    values ( \
    "0.070, 0.080", "0.071, 0.078", \
    "0.071, 0.081", "0.073, 0.079", \
    "0.066, 0.082", "0.068, 0.085");
  }
}
```

Just like the case of combinational cells, the switching power can be dissipated even
when the outputs or the internal state(s) does not have a transition. A common
example is the clock that toggles at the clock pin of a flip-flop. Significant power is
dissipated in the flip-flop with each clock toggle even if the flip-flop does not change
state. This is typically due to switching of an inverter inside of the flip-flop cell. An
example of the input clock pin power specification is shown below.

```
cell (DFF) {
  . . .
  pin (CLK) {
    . . .
    internal_power () {
    when : "(D&Q) | (!D&!Q)";/* No output
                                transition */
      rise_power (template_3x1) {
        index_1 ("0.1, 0.25, 0.4");/* Input
                                      transition */
        values ( /* 0.1       0.25       0.4 */ \
                  "0.045,     0.050,     0.090");
    }
      fall_power (template_3x1) {/* Inactive clock
            edge only, internal inverter switching */
        index_1 ("0.1,    0.25,   0.4");
        values ("0.045, 0.050, 0.090");
    }
    }
    internal_power () {
    when : "(D&!Q) | (!D&Q)"; /* Output switching
                                on active clock edge */
      rise_power (scalar) { /*Input switching power
    included with internal power tables for output */
        values ( "0" );
    }
      fall_power (template_3x1) { /* Inactive clock
            edge only, internal inverter switching */
        index_1 ("0.1,    0.25,   0.4");
        values ("0.045, 0.050, 0.090");
    }
    }
  }
}
```

This example shows the power specification for the *CLK* pin toggles described within two *when*-conditions.

The first *when*-condition ("(D&Q)|(!D&!Q)") represents the case when the output Q is the same as D and thus the clock does not cause any change in the state of the flip-flop. The internal power in this case is essentially the power due to clock inverter (within the flip-flop) switching when the clock switches.

The second *when*-condition ("(D&!Q)|(!D&Q)") represents the case when the clock results in a change in state of the flip-flop. The internal power in this case is only for the inactive clock edge (in this case, the falling edge of the clock). The power due to active (or rising) edge of the clock is modeled with the output pin Q switching due to the *CK*.

2.2.3   Internal Power Dependence of Parameters

The internal power is primarily dependent upon the power supply and there is roughly a quadratic dependence with respect to the power supply voltage. The dependence with respect to other parameters can be summarized as:

(a) *Threshold Vt of cells*: The high Vt^2 (HVt) cells have a lower internal power with respect to the low Vt (LVt) cells. For a specific technology node, up to 20% reduction in internal power is possible by changing to an equivalent higher Vt cell.

(b) *Process corner*: The internal power is higher for the *fast* process corners. For the same power supply and temperature, the internal power for *fast* process can be ~10% higher than at the *nominal* process condition.

(c) *Temperature*: The internal power normally increases with temperature due to slower transition rates of the signals at higher temperature. This increase with temperature, while negligible for slow process, high Vt cells and low power supply voltage, is more significant for fast process, low Vt cells and high power supply voltage. Under typical conditions (typical process, standard Vt and nominal power supply voltage), the internal power dissipation at higher temperature (such as 125°C) can be ~10% higher than the internal power dissipation at low temperature (such as −40°C).

(d) *Channel length*: Similar to the Vt, cells using *longer* channel devices have lower internal power than the cells using *shorter* channel devices.

Note that in addition to the parameters listed above, there are variation across cells in a library—with higher drive strength cells having larger internal power than the weaker cells. Since a design uses a variety of cell types, we focus on parameters listed above to describe the variation of internal power for a design.

Overall for a given design, the internal power is, in general, highest at the *fast* process, maximum power supply, and maximum temperature. Thus, this corner is typically used for specifying the worst power of the device.

2.3   Leakage Power

The standard cells and memory macros in CMOS digital logic are designed with the intent that the power is dissipated only when there is activity at the pins of the standard cells or the memory macros. However, as described in Sect. 2.1, power is dissipated even when there is no activity in the design. This is due to leakage in the MOS devices and this contribution is called the *leakage power*.

---

[2] The high Vt cells refer to cells with higher threshold voltage than the standard for the process technology.

Table 2.1 Cell delay and leakage power for different thresholds in 32 nm process

Threshold Vt	Cell delay (ps)	Leakage power (nW)
Ultra high Vt (uHVt)	46.9	0.146
High Vt (HVt)	36.7	0.677
Standard Vt (SVt)	26.4	1.99
Low Vt (LVt)	18.9	12.15
Ultra low Vt (uLVt)	17.1	41.2

2.3.1 Dependence on Threshold Voltage

As described in Sect. 2.1, the leakage power contribution is mainly from two phenomena: subthreshold current in the MOS device and gate oxide tunneling. By using high Vt cells, one can reduce the subthreshold current; however, there is a trade-off due to the reduced speed of the high Vt cells. The high Vt cells have a smaller leakage but are slower in speed. Similarly, the low Vt cells have a larger leakage but provide greater speed. The contribution due to gate oxide tunneling does not change significantly by switching to high (or low) Vt cells.

Based upon the above, a possible way to control the leakage power is to utilize high Vt cells. Similar to the selection between high Vt and standard Vt[3] cells, the strength of cells used in the design provides a trade-off between leakage and speed. The higher strength cells have higher leakage power but provide higher speed. The trade-offs related to power management are described in detail in later chapters.

As an example, Table 2.1 shows the leakage power and the cell delay[4] at the typical corner for an inverter cell using various Vt in a 32 nm process.

2.3.2 Dependence on Channel Length

In general, most standard cell libraries are built with the minimum channel length supported by the technology. However, the leakage current can be reduced by increasing the channel length of the MOS devices. For example, in 40 nm process technology, 40 nm is the smallest channel length possible for building the core devices. The leakage in the standard cells can be reduced by building standard cells using a longer channel length. For the example of the 40 nm process technology, the standard cells can be built using 45 nm or even 50 nm channel lengths. The longer channel devices would have lower performance (larger delays); however the leakage in these cells is lower.

---

[3] Standard Vt (SVt) is also sometimes referred to as Regular Vt (RVt).

[4] Using representative input slew and output loading.

Table 2.2 Cell delay and leakage power for different channel lengths in a 40 nm process

Channel length (nm)	Delay (ps)	Leakage power (nW)
40 nm	26.2	48.4
50 nm	31.8	24.4

Table 2.3 Leakage power variation with temperature for a representative cell in 45 nm low power process

Temperature (°C)	Leakage power (nW)
25	0.072
105	1.426
125	2.613

The standard cells built with longer channel devices provide a choice for trading off performance with leakage without changing the threshold of the devices. As an example, Table 2.2 shows the trade-off between leakage power (for *typical* process at room temperature) and delays for a buffer cell in a 40 nm process.

2.3.3 Dependence on Temperature

The subthreshold MOS leakage has a strong non-linear dependence with respect to temperature. In most process technologies, the subthreshold leakage grows by 10–20× or even higher as the device junction temperature is increased from 25°C to 125°C. The contribution due to gate oxide tunneling has a much smaller variation with respect to temperature or the Vt of the devices. The gate oxide tunneling, which was negligible at process technologies 100 nm and above, is a significant contributor to leakage at lower temperatures for 65 nm or finer process technologies. For example, gate oxide tunneling leakage may exceed the subthreshold leakage at room temperature for 65 nm or finer process technologies. At high temperatures, the subthreshold leakage continues to be the dominant contributor to leakage power. Table 2.3 shows the leakage of a typical 2-input *nand* cell which increases very rapidly as the temperature is increased.

2.3.4 Dependence on Process

The leakage power has a strong dependence on the process corner—the leakage for the *fast* process corner is much higher than the leakage at the *typical* process. Similarly, the leakage for the *slow* process corner is much smaller than the *typical* leakage. Table 2.4 shows the leakage of a 2-input *nand* cell for a 45 nm low power process.

Table 2.4 Process vs leakage

Process	Leakage (nW)
Slow	0.06
Typical	0.58
Fast	10.7

2.3.5   Modeling of Leakage Power

Leakage power for a standard cell is specified for each cell in the library. For example, an inverter cell may contain the following specification:

```
cell_leakage_power : 1.366;
```

This is the leakage power dissipated in the cell. The leakage power units are as specified in the header of the library, typically in nanowatts (nW). In general, the leakage power depends upon the state of the cell and state-dependent values can be specified using the *when*-condition. For example, an *inverter* cell can have the following specification:

```
cell_leakage_power : 0.70;

leakage_power() {
  when : "!I";
  value : 1.17;
}
leakage_power() {
  when : "I";
  value : 0.23;
}
```

where I is the input pin of the *inverter* cell. It should be noted that the specification includes a default value (outside of the *when*-conditions) and that the default value is generally the average of the leakage values specified within the *when*-conditions.

In a more general standard cell, the leakage power is provided for various possible states of the cell. Thus, if the state of the CMOS design is known, the leakage can be computed based upon the state of each cell or macro in design.

2.4   Advanced Power Modeling

The modeling of dynamic power in Sect. 2.2 is using the traditional table (or nonlinear) models described in terms of input transition time and the output capacitance. More accurate models of the dynamic behavior is available using current based models—such as the Composite Current Source (CCS) models or the Effective Current Source Models (ECSM). Just as the current based CCS or the ECSM models provide greater accuracy for delay calculation in presence of the interconnect, these models also provide greater flexibility in the modeling of power.

The current based models specify the current information (i.e. power supply current for the leakage as well as the power supply transient current during switching). This allows the current based models to be usable for dynamic simulations also. Note that the power models described in Sect. 2.2 describe the total energy dissipated in a transition and thus are not usable for detailed time-domain simulations of the power supply network. The details of the CCS power models are described next.

2.4.1 Leakage Current

This is analogous to the leakage power as described in Sect. 2.3. Instead of the leakage power, the model specifies the leakage current from the power supply. In addition to the power supply leakage current, gate leakage current is also specified by the model. An example excerpt is shown below.

```
leakage_current () {
  when : "A1 !A2 ZN";
  pg_current (VDD) {
    value : 6e-6;
  }
  pg_current (VSS) {
    value : -4e-6;
  }
  gate_leakage (A1) {
    input_high_value : 8e-9;
  }
  gate_leakage (A2) {
    input_low_value : -2e-6;
  }
}
```

The above example of a 2-input *nand* cell specifies the leakage currents from the power supply pins as well as the (gate) leakage currents through the input pins *A1* and *A2* of the cell. The negative values of current above imply that the current flows out of the cell *VSS* pin and that the gate leakage current flows out of the cell input pin *A2* (when *A2* is *low*). This example shows state dependent leakage for the specific condition of the inputs *A1* (= logic-1) and *A2* (= logic-0). The other state dependent leakage currents are specified in a similar manner.

2.4.2 Dynamic Current

This is analogous to the active power described in Sect. 2.2. The power supply current is specified for different combinations of input transition time and output capacitance. For each of these combinations, the power supply current waveform is specified. Essentially, the waveform here refers to the power supply current values specified as a function of time. An example of the power supply current for an output rise condition is shown next.

```
dynamic_current() {
  related_inputs : "A";
  related_outputs : "ZN";
  switching_group () {
    input_switching_condition (fall);
    output_switching_condition (rise);
    pg_current (VDD) {
      vector ("ccs_power_template_1") {
        reference_time : 5.06; /* Time of input
                                  crossing threshold */
        index_1 ("0.040"); /* Input transition */
        index_2 ("0.900"); /* Output capacitance */
        index_3 ("5.079e+00, 5.093e+00, 5.152e+00,
           5.170e+00, 5.352e+00");/* Time values */
        /* Power supply current: */
        values ("9.84e-06, 0.082, 0.081,
                 0.157, 0.0076");
      }
      . . .
    }
    . . .
  }
  . . .
}
```

The *reference_time* attribute refers to the time when the input waveform crosses the input threshold. The *index_1* and *index_2* refer to the input transition time and the output load used and *index_3* is the time. The *index_1* and *index_2* (the input transition time and output capacitance) can have only one value each. The *index_3* refers to the time values and the table values refer to the corresponding power supply current. Thus, for the given input transition time and output load, the power supply current waveform as a function of time is available. Additional lookup tables for other combinations of input transition time and output capacitance are also specified.

Power supply current for other scenarios are also described similarly. An example of the power supply current due to transition at the input which does not result in an output transition is described below:

```
dynamic_current() {
  related_inputs : "A" ;
  when : "!B & ZN" ;
  switching_group () {
    input_switching_condition (rise);
```

```
pg_current (VDD) {
  vector ("ccs_power_template_2") {
    reference_time : 5.06; /* Time of input
                             crossing threshold */
    index_1("0.040"); /* Input transition */
    index_2("5.079e+00, 5.093e+00, 5.152e+00,
        5.170e+00, 5.352e+00");/* Time values */
    /* Power supply current: */
    values("6.44e-06, -0.012, 0.008,
            0.003, 1.09e-04");
  }
  . . .
  }
  . . .
 }
 . . .
}
```

One major advantage of using current based models is that these are usable for dynamic power supply simulations to estimate power supply transient noise based upon the activity and the decoupling capacitances present in the system.

2.5 Summary

The leakage power is dependent only on the state of nets in the design (*logic-0* or *logic-1*) whereas the active power is dependent upon the switching activity in the design.

The activity can be at the input or at the output pins of the cells. The power dissipation is a function of the following characteristics of the design:

- Types of cells used for implementation—this affects both active and leakage power.
- External environment such as power supplies and temperature.
- Activity within the design and operating frequency (or frequencies).
- IO interfaces and external loading.
- Characteristics of embedded memory macros.

The power models in the standard cell library are utilized to obtain the power dissipation in the design arising from core logic, that is, for the portion of the design implemented using standard cells.

The next chapter describes the power dissipation in memory macros, IOs and other macros (such as analog blocks) which may be used in the CMOS digital design.

Chapter 3
Modeling of Power in IOs and Macro Blocks

This chapter describes various aspects of power dissipation in macro blocks and IOs[1] in a CMOS design. As described in Chap. 2, the memory macros and the IOs can have power dissipation due to switching activity, called active power, and another contribution called leakage power that represents the power without any activity in the macro. This chapter describes both of these contributions specifically for the macro blocks (analog blocks, memory macros) and IOs in the design.

3.1 Memory Macros

Embedded memory macros are significant contributors to the power dissipation within an ASIC. Figure 3.1 shows a simplified black-box representation of a single port memory macro.

The figure shows two input buses and one output bus. The *DATA* and *ADDR* buses are input buses and *Q* is an output bus. In addition, there are three other inputs: a clock input (*CLK*) and two enable inputs—memory enable (*ME*) and write enable (*WE*). At the positive clock edge, the memory macro is in *write* mode when both memory enable and write enable are active, else the memory macro is in *read* mode when only memory enable is active. If memory enable is inactive, nothing useful happens at the positive clock edge.

With the above overview, we can describe the power dissipation for the memory macro. The active and the leakage power dissipation for the memory macros are described in the following two subsections.

---

[1] Input, output or bidirectional buffers.

R. Chadha and J. Bhasker, *An ASIC Low Power Primer: Analysis,*
Techniques and Specification, DOI 10.1007/978-1-4614-4271-4_3,
© Springer Science+Business Media New York 2013

Fig. 3.1 Single port
memory macro

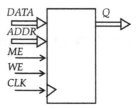

3.1.1   *Dynamic or Active Power*

For each of the input buses, such as *ADDR* and *DATA*, the internal power is dissipated for each transition of the signals in these buses. For example, the *ADDR* and
DATA buses cause internal power dissipation in the memory whenever there is
switching activity at the inputs. An example of an internal power table in a memory
macro for the address bus is as illustrated below.

```
bus(ADDR) {
  bus_type : ADDR_9_0;
  direction : input;
  pin(ADR[9:0]) {
   /* Address power */
   internal_power() {
    rise_power(scalar) {
     values ("0.124");
    }
    fall_power(scalar) {
     values ("0.124");
    }
   }
  }
}
```

The above fragment illustrates the internal power dissipation due to *ADDR* bus
activity. The values above are internal power dissipation for each bit of the *ADDR*
bus. The power values above can also be modeled as dependent upon the state of the
control pins such as memory enable pin *ME*.

The fragment below illustrates the input power dissipation due to switching of
the *DATA* bus.

```
bus(DATA) {
  bus_type : DATA_31_0;
  direction : input;
  pin(DATA[31:0]) {
   /* Data power */
```

```
    internal_power() {
     rise_power(scalar) {
      values ("0.153");
      }
     fall_power(scalar) {
      values ("0.153");
      }
     }
    }
   }
```

Similar internal power dissipation occurs for the control signals—write enable (*WE*) and memory enable (*ME*) pins—as shown below.

```
  pin(WE) {
    direction : input;
    /* Write enable power */
    internal_power() {
     rise_power(INPUT_BY_TRANS) {
      values ("17.08, 17.08, 17.08");
      }
     fall_power(INPUT_BY_TRANS) {
      values ("17.08, 17.08, 17.08");
      }
     }
   }
  pin(ME) {
    direction : input;
    /* Memory enable power */
    internal_power() {
     rise_power(INPUT_BY_TRANS) {
      values ("0.048, 0.048, 0.048");
      }
     fall_power(INPUT_BY_TRANS) {
      values ("0.048, 0.048, 0.048");
      }
     }
   }
```

The values for power in the example above are representative and illustrate that the switching of the write enable (*WE*) signal results in a much higher internal power dissipation as compared to the switching of the memory enable (*ME*) signal or even switching of the *DATA* or *ADDR* bus signals.

A rising transition at the clock pin, with appropriate values at the memory enable and the write enable pins, results in a *read* or a *write* operation; this is the dominant cause for power dissipation in the memory macro.

```
pin(CLK) {
  /* Write power */
  internal_power() {
   when : "ME & WE";
   rise_power(INPUT_BY_TRANS) {
    values ("42.4, 42.4, 42.4");
   }
   fall_power(INPUT_BY_TRANS) {
    values ("0.0, 0.0, 0.0");
   }
  }
  /* Read power */
  internal_power() {
   when : "ME & !WE ";
   rise_power(INPUT_BY_TRANS) {
    values ("43.9, 43.9, 43.9");
   }
   fall_power(INPUT_BY_TRANS) {
    values ("0.0, 0.0, 0.0");
   }
  }
  /* Disabled power */
  internal_power() {
   when : "!ME";
   rise_power(INPUT_BY_TRANS) {
    values ("0.93, 0.93, 0.93");
   }
   fall_power(INPUT_BY_TRANS) {
    values ("0.0, 0.0, 0.0");
   }
  }
}
```

The power dissipation tables illustrate that the clock input transitions for the memory *read* or *write* operations result in a much higher power dissipation in the memory macro than the activity of other inputs.

During the *read* operation, the output bus Q also switches which results in internal switching as well as output charging power. The internal switching power for the output bus Q is depicted next.

```
bus(Q) {
  bus_type : Q_31_0;
  direction : output;
  pin(Q[31:0]) {
   /* Dataout power */
   internal_power() {
    rise_power(scalar) {
     values ("0.022");
    }
    fall_power(scalar) {
     values ("0.022");
    }
   }
  }
}
```

Note that the output charging power for the output Q bus would depend upon the output capacitive load driven by the Q pins of the memory. However, in most practical cases, the dynamic power dissipation in the memory is governed by the *read* and *write* performed by the clock. The active power computation of a memory instance with representative activity at its pins is described in Chap. 4.

The next section describes the leakage power modeling for the SRAM macros.

3.1.2 Leakage Power

Unlike standard cells where the leakage power description is normally state-dependent, the leakage power for memories is usually not described as dependent upon address, data or clock inputs.[2] As an example, the leakage power for a memory macro can be represented as:

```
cell_leakage_power : 1622700;
```

The memory macros can be thought of as comprised of two portions: the core memory array (which stores the memory information), and the peripheral logic which is comprised of the address decoders, bit-line prechargers, sense amplifiers, and other driver circuitry, etc. Many memory implementations provide separate power supply connections for the memory core and memory peripheral logic.In such cases, the leakage power for the memory macro is described for each of the two power supplies. An example of the leakage power specification for the memory macro with separate power supplies is shown below:

---

[2] The leakage power of the memory macro does depend upon various control signals which may place the memory in one of the *shutdown* or the *low leakage* modes.

```
leakage_power() {
  related_pg_pin : "VDDPE";
  value : 1135800;
}
leakage_power() {
  related_pg_pin : "VDDCE";
  value : 486900;
}
```

In above, *VDDPE* is the power supply for the memory peripheral logic and the *VDDCE* is the power supply for the memory core array (where the memory contents are stored). In this example, the peripheral logic accounts for the larger proportion of the leakage power. This is generally true for smaller memory macros. The memory core array leakage would be proportionally larger for larger memory instances.

3.1.2.1   Trade-Off Between Leakage Power and Speed

The memory core array bit cells are generally provided by the foundry or built by the memory macro provider. For the peripheral logic, there can be a choice of selecting the threshold (Vt) of the MOS transistors. For example, peripheral logic built with low Vt MOS transistors would provide higher speed for the memory macro—however the leakage power would be higher due to low Vt transistors used. The trade-offs between leakage power and speed, obtained by selecting different threshold transistors in the peripheral logic, is similar to the standard cell logic trade-off described in Sect. 2.3.

Similarly, the peripheral logic can use longer channel MOS transistors to reduce leakage power with some slowdown in memory access time. This is also similar to the standard cell logic trade-off described in Sect. 2.3.

The above trade-off between leakage power and speed are applicable during normal memory operation. Similarly, when the memory is inactive, certain techniques can be adopted to reduce the leakage power for the memories. These methods are employed when the memory macro is inactive and thus there is no trade-off with respect to performance.

3.1.2.2   Controlling Leakage Power in Inactive Mode

For a memory macro in inactive mode, the following are a few of the techniques used to reduce the leakage power:

(a) *Shutting down of peripheral logic*. The peripheral logic block can be shut down. The control for shutting down of the peripheral power supply can be external or internal to the memory macro. Shutting down the peripheral supply essentially eliminates the leakage contribution from the peripheral logic. This method of shutting down peripheral logic may be termed as *light sleep* (or *snooze*) mode by some memory providers.

(b) *Back bias for core memory array.* Since the core memory array stores the memory contents, shutting down the core array supply would imply loss of the memory content. Thus, the leakage from the core memory array can only be reduced by adding a back bias for the memory array logic. The back bias reduces the leakage contribution significantly while retaining the contents of the memory. This method of adding *back bias* to the core array may be referred to as *deep sleep* mode by some memory providers.

Either or both of the methods can be utilized for leakage reduction. The key is that these methods can only be employed when the memory macro access is not required for a significant duration to justify the savings. After the design sets the control signals for placing the memory macro in the *light sleep* or the *deep sleep* mode, there is certain delay before the leakage reductions for the *light sleep* or the *deep sleep* mode are achieved. Similarly there is a delay in bringing it back from the *sleep* modes to *normal* mode.

Each of the techniques for reducing the leakage power requires adding other control signals to enable the low power modes. As an example, for a memory macro with a light sleep (*LS*) input, the leakage power is described as:

```
leakage_power () {
  when   :  "!LS";
  value  :  1622700;
}
leakage_power () {
  when   :  "LS";
  value  :  777000;
}
```

The above example is for a memory macro with one power supply for the core array and the peripheral logic.

3.2 Power Dissipation in Analog Macros

The power dissipation in the analog macros, such as PLLs and SerDes macros, normally cannot be classified into leakage and active power components. For example, these macros may dissipate significant DC power in the voltage-controlled oscillator (VCO) block. The incremental power due to output frequency is normally a smaller contribution than the baseline DC power. The power dissipation in analog macros is thus described as DC power plus a variable contribution, such as due to output frequency.

The power dissipation for the analog blocks is generally specified only in a limited form, for example, as line items in the datasheet. Here is an example from a SerDes macro specification.

Normal Power dissipation (at 6 Gbps)
AVDD: 10.5mA
AHVDD: 20mA
VDD_CORE: 20μA

The power dissipation normally depends upon various modes of operation. Examples of such modes are: low speed mode, high speed mode and power down mode.

3.3   Power Dissipation in IO Buffers

In general, a significant portion of the power in a design is dissipated in the IO buffers. This is largely because in comparison to the core signals, the IO signals have a larger voltage swing and the outputs drive a large capacitive load.

3.3.1   General Purpose Digital IOs

For general purpose digital IOs, the power computation is largely similar to that of the standard cell logic. The digital IOs can be broadly considered as one of the following:

(a) Output IO buffers
(b) Input IO buffers
(c) Bidirectional IO buffers

The power computation for each of the above scenarios is described next.

3.3.1.1   Output IO Buffers

For the output IO buffer, the power dissipation is obtained from:

(a) The leakage power
(b) The internal switching power, and
(c) The output charging power

The leakage power for the IOs is usually a small portion of the total power. (This is different from the case of the standard cell logic where leakage can be a significant portion of the total power.) Amongst the other two power components (internal switching power and output charging power), the power due to the output load (that is external) of the IOs is normally much larger than the internal switching power. An excerpt of the leakage power specification for an IO buffer is given below.

```
rail_connection(PV1, CORE_VOLTAGE);
rail_connection(PV2, IO_VOLTAGE);
cell_leakage_power : 940.31;
leakage_power() {
  power_level : "CORE_VOLTAGE";
  when : "!I";
  value : 866.56;
}
leakage_power() {
  power_level : "IO_VOLTAGE";
  when : "!I";
  value : 83.92;
}
leakage_power() {
  power_level : "CORE_VOLTAGE";
  when : "I";
  value : 848.44;
}
leakage_power() {
  power_level : "IO_VOLTAGE";
  when : "I";
  value : 81.70;
}
```

An IO buffer also acts as a level shifter since the signal level external to the ASIC is generally different from the signal level inside the core of the ASIC. Thus, an IO buffer uses two power supplies—an *IO_VOLTAGE* and a *CORE_VOLTAGE*. The leakage power for multi-voltage cells needs to be associated with each power supply. The above excerpt for leakage power in an IO buffer shows the leakage power for *IO_VOLTAGE* and *CORE_VOLTAGE* separately. The leakage power values for both the core and the IO supplies are dependent upon the state of the input pins of the IO buffer.

The excerpt below depicts the internal power models for the IO buffer.

```
internal_power(){
  power_level : IO_VOLTAGE;
  related_pin : "I";
  rise_power(TABLE_LOAD_2x3) {
    index_1 ("0.1, 0.4"); /* Input transition */
    index_2 ("4.0, 8.0, 12.0"); /* Output cap /
    values("12.84, 12.87, 12.89", \
        "12.85, 12.88, 12.90");
  }
```

Fig. 3.2 Output IO buffer

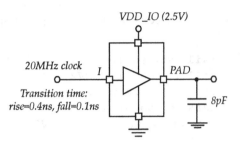

```
fall_power(TABLE_LOAD_2x3) {
  index_1 ("0.1, 0.4"); /* Input transition */
  index_2 ("4.0, 8.0, 12.0"); /* Output cap */
  values("11.01, 11.04, 11.11", \
         "10.99, 11.02, 11.06");
}
```

The power model above is related to the *IO_VOLTAGE*. In this example, no separate power models are specified for *CORE_VOLTAGE* since the power dissipated from core power supply is negligible.

Example Computation of Active Power. Consider an example of an output IO buffer in Fig. 3.2. The buffer operates at an output power supply (*VDD_IO*) of 2.5 V and drives an output load of 8 pF. The input pin *I* has a clock with frequency 20 MHz applied to it. The clock signal at pin *I* has a rise transition time of 0.4 ns and a fall transition time of 0.1 ns. What is the total active (or dynamic) power dissipated in the output buffer for the *VDD_IO* supply?

Based upon the internal power tables provided:

Energy dissipated per rise transition
(contributing to internal switching power) =
12.88pJ per transition

Energy dissipated per fall transition
(contributing to internal switching power) =
11.04pJ per transition

Internal switching power =
(12.88 + 11.04) * (1E-12) * 20 * 1E6 =
478.4μW

Output charging power (CV$_{dd}^2$f) =
8.0 *(1E-12)* 2.5 * 2.5 * (20 * 1E6) =
1000μW

Total active power (internal plus output) = **1478.4μW**

Fig. 3.3 Input IO buffer

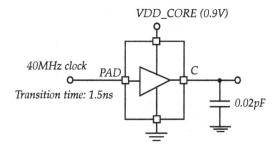

A total of 1478.4 μW of active power is dissipated within this output buffer from the *VDD_IO* supply. Because of larger output voltage and higher output loading, the active power of the output buffer is dominated by the output charging power.

3.3.1.2   Input IO Buffers

For the input IO buffer, the overall breakdown of power is the same as for output IO buffers. One difference is that the output charging now refers to the core side pin of the IO and thus the output charging contribution is similar to that of the other core logic signals. Here is an example excerpt from the Liberty description of an input IO buffer.

```
internal_power(){
  power_level  :  CORE_VOLTAGE;
  related_pin  :  "PAD";
  rise_power(TABLE_LOAD_2x3) {
    index_1 ("1.0, 2.0"); /* Input transition */
    index_2 ("0.01, 0.02, 0.04"); /* Output cap */
    values("1.12, 1.12, 1.12", \
           "0.84, 0.84, 0.85");
  }
  fall_power(TABLE_LOAD_2x3) {
    index_1 ("1.0, 2.0"); /* Input transition */
    index_2 ("0.01, 0.02, 0.04"); /* Output cap */
    values("6.43, 6.44, 7.02", \
           "5.29, 5.32, 5.64");
  }
```

The power model above is related to the *CORE_VOLTAGE*. In this example, no separate power models are specified for *IO_VOLTAGE* since the power dissipated from the IO power supply is considered to be negligible.

Example Computation of Active Power. Consider an example of the input IO buffer in Fig. 3.3. The buffer operates at the core power supply of 0.9 V and drives a

0.02 pF load at pin *C*. The input pin *PAD* has a clock with a frequency of 40 MHz applied to it and both rise and fall transition times at the *PAD* pin are 1.5 ns. What is the total active (or dynamic) power dissipated in the input buffer for the core power supply?

Based upon the internal power table above:

```
Energy dissipated per rise transition (contributing to
internal switching power) = (1.12 + 0.84)/2
= 0.98pJ per transition
```

```
Energy dissipated per fall transition (contributing to
internal switching power) = (6.44 + 5.32)/2
= 5.88pJ per transition
```

```
Internal switching power =
(0.98 + 5.88) * 40 * 1E6 =
274.4µW
```

```
Output charging power (CVdd²f) =
0.02 *(1E-12)* 0.9 * 0.9 * (40 * 1E6) =
0.648µW
```

```
Total active power (internal plus output) = 275.048µW
```

Unlike the output buffer, the output charging has a very small contribution to the total active power.

3.3.1.3 Bidirectional IO Buffers

The bidirectional IOs can operate either in input mode or in output mode. Thus, based upon a control signal, the IO operates as an input IO buffer or as an output IO buffer. In the input mode, the input signal at the *PAD* is propagated to the core side pin *C*, and in the output mode, the signal at the core side input pin *I* is transferred to the output *PAD*. Note that the *PAD* signal is available at the core side pin *C* even in the output mode. Since the *PAD* signal is always available at the core side pin *C*, the power corresponding to the input mode is dissipated even when the buffer is in output mode.

The average power dissipation for a bidirectional IO buffer is thus obtained by combining the power dissipations from the portion of time the IO buffer is in input mode with the power dissipation for the portion of time the IO buffer is in output mode. This assumes that the toggle rate for the output enable control signal is negligible in comparison to the toggle rates of the signals at pin *I* or at *PAD*.

Example Computation of Active Power for Bidirectional IO. Consider an example where the bidirectional buffer has the input mode and output mode power models same as those for the input and output buffers described earlier.

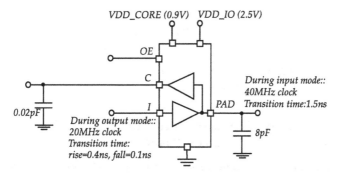

Fig. 3.4 Bidirectional IO buffer

The bidirectional buffer shown in Fig. 3.4 acts in the input mode for 60% of the time and is in the output mode for the remaining 40% of the time. For input mode, the *PAD* signal has a 1.5 ns transition time for both rise and fall transitions. For output mode, the core side input pin (*I*) has transition times of 0.4 ns for rise and 0.1 ns for fall transitions. The load on the *PAD* pin is 8 pF whereas the load on the core side output pin (*C*) is 0.02 pF. During the input mode, a clock with a frequency of 40 MHz is applied at the *PAD* and during the output mode, a clock with a frequency of 20 MHz is applied at the core side pin (*I*). We assume that the output transition times at *PAD* are 1.5 ns for both the rise and the fall transitions (same as the transition times for the externally applied *PAD* signal in the input mode). What is the total active power assuming a core voltage of 0.9 V and an IO voltage of 2.5 V?

The internal power tables for the output mode and the input mode are the same as the internal power tables for the output buffer and the input buffer described previously in this section. In addition, the activity (clock frequency), input transition times and output loading values are also the same as the values assumed for the output and input buffers in this section. Based upon the previous example computations:

Power for I->PAD transitions during output mode = 1478.4μW
(same as for Output IO buffer)

For *PAD->C* transitions in the output mode, the following computation is used. This is similar to the power calculation for the input IO buffer.

```
Energy dissipated per rise transition (contributing to
internal switching power) = (1.12 + 0.84)/2
= 0.98pJ per transition

Energy dissipated per fall transition (contributing to
internal switching power) = (6.44 + 5.32)/2 =
5.88pJ per transition

Internal switching power =
(0.98 + 5.88) * 20 * 1E6 =
137.2μW
```

```
Output charging power (CV_dd²f) =
0.02 *(1E-12)* 0.9 * 0.9 * (20 * 1E6) =
0.324µW
```

Power for PAD->C transitions during output mode (internal plus output) = 137.2 + 0.324 =
137.524µW

Power during Output mode (40% of time)=
1478.4 + 137.524 = **1615.924µW**

Power during Input mode (60% of time)= 275.048µW (same as for Input IO buffer)

Total active power for bidirectional buffer =
40% of 1615.924 + 60% of 275.048 =
646.3696 + 165.0288 =
811.3984µW

3.3.2   High Speed IOs with Termination

For high speed interfaces, the IOs are normally terminated to minimize reflections on the line. In such cases, the swing on the transmission line is not rail-to-rail but is reduced due to parallel resistive termination. In addition, the power dissipation has a DC component due to the resistive termination.

3.3.2.1   Output Buffer with External Parallel Termination

Figure 3.5 depicts an example of the output IO with external parallel termination. (Note that the same representation applies to a bidirectional IO buffer in the output mode.) For a point-to-point connection, the board level interconnect trace is designed as a transmission line with a specific impedance. An impedance mismatch at the two ends (of the transmission line) creates reflections which are not desirable (especially for high speed interfaces). Thus, it is common to have parallel termination at the destination to minimize reflections. The termination *Rt* (Fig. 3.5) is connected to a *VTT* supply which is normally set to *VDDQ/2* (where *VDDQ* is the output voltage supply of the IO buffer).

While the termination *Rt* is depicted to be outside of the destination macro in Fig. 3.5, many cases such as DRAMs have built-in termination internal to the macro. This is also referred to as input ODT (On-Die Termination).

Even in steady high (or steady low) state, the output buffer depicted in Fig. 3.5 has a DC current through the output stage of the IO buffer. When the IO buffer is output *high*, the DC current flows from the output *PAD* pin through the transmission line into the termination resistor *Rt* into the *VTT* supply. Similarly, when the IO buf-

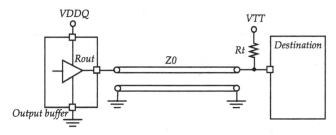

Fig. 3.5 *Rt* external termination (to *VTT*) at far end of transmission line

Table 3.1 Output voltage at *PAD* pin of IO with and without termination

	Vout (V)—with 66 ohm (Rt) termination to VTT (=VDDQ/2)	Vout (V)—with no termination
Output high	1.245	1.5
Output low	0.255	0

fer is output *low*, the DC current flows from the termination voltage supply *VTT* through the transmission line into the *PAD* pin of the output buffer to ground supply. Assume an IO with an *Rout* of 34 ohm,[3] and a 66 ohm termination impedance at destination. With 1.5 V for *VDDQ* and 0.75 V for *VTT*, the voltage levels at the *PAD* pin are depicted in Table 3.1.

DC power dissipated: In the above IO example, the DC power dissipated in the IO is:

```
DC power in IO = 0 (if no termination)⁴

DC power in IO =
  (0.255 * 0.255) / 34 =
  1.913mW (with 66ohm termination to VDDQ/2)
```

Note that the above computation is for power dissipated within the IO buffer; there is additional power (supplied by *VDDQ* of the IO buffer) which is dissipated in the 66 ohm termination resistor.

Power supplied by VDDQ *of the IO buffer*: The current and the power supplied by *VDDQ* of the IO buffer is depicted in Table 3.2. The table illustrates that the power sourced by the *VDDQ* of the IO buffer is a different computation from the power dissipated within the IO buffer.

Just like in the case of analog macros, the power description in the Liberty models is often not adequate for accurate power analysis in presence of resistive termination.

---

[3] 34 ohms is one of the JEDEC standard drive impedance for DDR3 [JED10].
[4] There is additional leakage power in the IO both with and without termination.

Table 3.2 Comparison between power dissipated within the IO and the power supplied by *VDDQ* of the IO

	Power dissipated within the IO buffer (mW)	Current from VDDQ supply of the IO (mA)	Power supplied by VDDQ of the IO (mW)
Output high	1.913	7.5	11.25
Output low	1.913	0	0

The dissipated power can be computed for steady state condition (output *high* or *low*) based upon the power supply *VDDQ*, termination power supply, output drive impedance of the IO and the termination resistance value. Detailed SPICE level analysis is generally required to obtain the power dissipated in the specific configuration based upon the frequency of operation and the activity within the IOs.

3.3.2.2   Input Buffer with Resistive Termination

Figure 3.6 depicts an example of the input IO with internal parallel termination (Note that the representation in Fig. 3.6 also applies to a bidirectional IO buffer in the input mode.). As described in the case of the output buffer with external parallel termination, the point-to-point connections on the board are made with the interconnect traces designed as transmission lines with a specific impedance. To minimize reflections due to impedance mismatch, a (programmable) resistive termination at the input pin is utilized. The termination *Rt* to *VDDQ/2* is implemented as two resistances of value '*2*Rt*' connected to the *VDDQ* supply and to *VSSQ* respectively (*VDDQ* and *VSSQ* are the power and ground supplies for the IO buffer).

Even in steady high (or steady low) state, the input mode IO depicted in Fig. 3.6 has DC current flowing through the input termination resistances of the IO buffer. When the input signal is *high*, the DC current flows from the source through the transmission line into the input pin of the IO. Similarly, when the input signal is *low*, the DC current flows from the input pin of IO through the transmission line into the source driver. For a driver with 40 ohm output drive[5] connected to an IO with 60 ohm[6] input termination and operating at *VDDQ* of 1.5 V, the voltage at the *PAD* pin is given in Table 3.3.

DC power dissipated: In the above IO example, the DC power dissipated in the IO is:

```
DC power in IO = 0 (if no termination)⁷

DC power in IO =
  [(1.2 * 1.2) + (0.3 * 0.3)] / 120 =
  12.75mW (with 60ohm input termination)
```

---

[5] 40 ohms is one of the JEDEC standard drive impedance for DDR3 [JED10].

[6] 60 ohms is one of the JEDEC standard input termination impedance for DDR3 [JED10].

[7] There is additional leakage power in the IO both with and without termination.

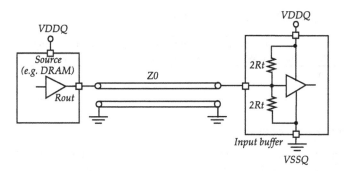

Fig. 3.6 Input buffer with input termination of *Rt*

Table 3.3 Input voltage at *PAD* pin of IO with and without termination

	Vin (V)—with input termination Rt of 60 ohm	Vin (V)—with no input termination
Source high	1.2	1.5
Source low	0.3	0

Note that the above computation holds for DC power dissipated within the IO and it applies when the source is *high* or when the source is *low*. Similar to the case described in previous subsection, the power sourced by the *VDDQ* supply of the IO buffer is an entirely different computation from the above.

Power supplied by VDDQ *of the IO buffer*: In the above IO example, the current and power supplied from the *VDDQ* of the IO are depicted in Table 3.4. Similar to the case described in previous subsection, the power dissipated in the IO is different from the power sourced by the *VDDQ* supply of the IO buffer.

Similar to the output mode, the power description in the Liberty models is normally not adequate for accurate power analysis in input mode also. The dissipated power can be computed for steady state condition (output *high* or *low*) based upon the power supply *VDDQ*, input termination resistance, and the output drive impedance of the driver. Detailed SPICE level analysis is generally required to obtain the power dissipated in the specific configuration based upon the frequency of operation and the activity within the IOs. In general, the input mode power is dominated by the DC power.

3.4 Summary

This chapter described the power computation for the memory macros, other core macro blocks and the IO buffers. Unlike the power for standard cell logic and memory macros, the power for special analog macros can have other dependencies (such as bias circuitry) which do not depend upon activity.

Table 3.4 Comparison between power dissipated within the IO and the power supplied by *VDDQ* of the IO

	Power dissipated within the IO buffer (mW)	Current from VDDQ supply of the IO (mA)	Power supplied by VDDQ of the IO (mW)
Source high	12.75	2.5	3.75
Source low	12.75	10	15

In summary, the key items are:

• The memory power is dependent upon whether the memory is enabled and also upon whether it is performing a read or a write operation.
• The memory leakage power can be reduced by placing the memory in one of the available sleep modes.
• The IO power is sourced from core as well as IO power supplies.
• The bidirectional IO power computation depends upon the portion of time the IO is in input mode versus the portion of time spent in output mode.
• The high speed IOs (e.g. such as DDR2/DDR3 IOs) generally use a terminated transmission line to reduce reflections. The parallel termination results in fixed DC power which is present even when the IO is not switching. Because of the parallel termination, the power dissipated within the IO buffer and the power supplied by the IO power supply are different.

The next chapter deals with detailed power computation based upon the activity in the design.

Chapter 4
Power Analysis in ASICs

This chapter describes various aspects of power analysis in a digital CMOS design. The power dissipation in an ASIC is comprised of power in the digital core logic, memories, analog macros, and other IO interfaces. The power dissipation in the digital logic and memory macros can be due to switching activity, called active power, and the leakage power which is present even with zero switching activity in the design. This chapter describes each of these contributions in detail—specifically the factors affecting the power calculation from various contributions in the design. Switching activity formats are also described in this chapter.

4.1 What Is Switching Activity?

As described in Chaps. 2 and 3, the power computation is generally obtained from the power models included in the library descriptions of the standard cells, memory macros and the IO libraries. This computation of power using library power models relies upon the transition activity and state of each pin of standard cells, memory macros and the IOs.

The key to power computation is the switching activity of each net. What is switching activity? The switching activity is comprised of the following two parameters:

(a) Static probability
(b) Transition rate

4.1.1 Static Probability

For a given net, the static probability refers to the expected state of the signal. For example, a static probability value of 0.2 implies that the signal is at logic-1 for 20%

R. Chadha and J. Bhasker, *An ASIC Low Power Primer: Analysis,*
Techniques and Specification, DOI 10.1007/978-1-4614-4271-4_4,
© Springer Science+Business Media New York 2013

of the time (and logic-0 for 80% of time[1]). A 50% duty cycle for a clock signal implies that the static probability of the clock signal is 0.5 (or the clock is logic-0 for 50% of time and logic-1 for 50% of time).

4.1.2 Transition Rate

The transition rate is the number of transitions per unit time. The transition rate is also referred to as toggle rate. For periodic signals such as clocks where the frequency of the signal is specified, the transition rate is twice the frequency of the signal (since there are two transitions—rising and falling—within each cycle).

The power analysis utilizes the switching activity (static probability and transition rate) for each signal in the design.

4.1.3 Examples

In Fig. 4.1, the probability that pins *CK* and *Q* are at 1 is 50%. However, the toggle rate for pin *CK* is 8 toggles in 40 ns, or 200 million transitions per second. The toggle rate for pin *Q* is 4 toggles in 40 ns, or 100 million transitions per second.

A net that has a probability of 1 or 0 is a constant net. A net with a probability of 0.5 is at logic-1 50% of time. This effectively describes the duty cycle of the net. A net with a probability of 0.25 is at logic-1 for 25% of time. An example of two waveforms with same toggle rate but different static probability values (different duty cycles) is shown in Fig. 4.2.

Consider the example shown in Fig. 4.3. Net *CK* has a probability of 0.5 and a toggle rate of 100 million transitions per second, and net *CKE* has a probability of 0.5 and a toggle rate of 2,000 transitions per second. In this case, the toggle rate for *CKG* is *nearly* 50 million transitions per second. This is because *CKE* has a probability of 0.5 (it is *on* for half the time) and the *CK* toggle rate is much larger than the *CKE* toggle rate. Thus, *CKE* can be treated as an *almost* steady signal in comparison, and one-half of the *CK* transitions would propagate to the *CKG* net.

4.2 Power Computation for Basic Cells and Macros

This section illustrates the detailed power computation of sample standard cells and memory macros using the library descriptions and the switching activity values.

---

[1] For the purposes of simplicity, we have assumed the signal cannot be in *unknown* (or X) state anytime.

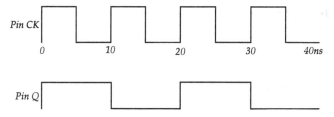

Fig. 4.1 Example waveforms with same static probability but different transition rates

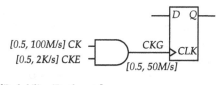

Fig. 4.2 Example waveforms with same transition rates but different static probability

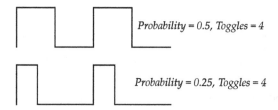

*[Probability, Toggle_rate]

Fig. 4.3 Example of reduced toggle rate at the output of an *and* gate

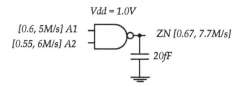

Fig. 4.4 NAND power computation using switching activity at the pins

4.2.1   Power Computation for a 2-Input NAND Cell

This section describes the power computation for a 2-input *nand* cell with input pins *A1* and *A2* and output pin *ZN*. Assume that the switching activity for the pins of the cell are available and are as shown in Fig. 4.4.

The power computation uses the switching activity along with the library description. A fragment of the power specification within the library for this cell is shown next.

```
leakage_power () {
  value : 42.2;
}
leakage_power () {
  value : 26.1;
  when : "!A1 !A2";
}
leakage_power () {
  value : 33.0;
  when : "!A1 A2";
}
leakage_power () {
  value : 27.0;
  when : "A1 !A2";
}
leakage_power () {
  value : 82.7;
  when : "A1 A2";
}
pin(A1) {
  direction : input;
  internal_power () {
   when : "!A2&ZN"; /* Transition at A1 does not */
                    /* cause an output transition. */
    rise_power (scalar) {
     values ("0.004");}
    fall_power (scalar) {
     values ("0.006");}
   }
  }
}
pin(A2) {
  direction : input;
  internal_power () {
   when : "!A1&ZN"; /* Transition at A2 does not */
                    /* cause output transition. */
    rise_power (scalar) {
     values ("0.006");
    }
    fall_power (scalar) {
     values ("0.008");
    }
   }
  }
}
```

```
pin(ZN) {
  direction : output;
  internal_power () { /* A1 causes output transition */
    related_pin : "A1";
    rise_power (scalar) {
      values ("0.043");}
    fall_power (scalar) {
      values ("0.016");}
  }
  internal_power () { /* A2 causes output transition */
    related_pin : "A2";
    rise_power (scalar) {
     values ("0.036");}
    fall_power (scalar) {
     values ("0.021");}
  }
}
```

The switching activity for the pins of the *nand* cell (with the library description for power described above) are shown in Fig. 4.4. These are normally obtained through simulation and the information is extracted in SAIF format. The switching activity values at the pins of the *nand* cell are:

```
Static probability (pin A1) = 0.6
Static probability (pin A2) = 0.55
Toggle rate (pin A1) = 5 million transitions/sec
Toggle rate (pin A2) = 6 million transitions/sec
Static probability (pin ZN) = 0.67
Toggle rate (pin ZN) = 7.7 million transitions/sec
```

Note that the static probability at the output ZN directly follows from the static probability values at the inputs $A1$ and $A2$.[2]

4.2.1.1   Leakage Power Computation

Leakage power is computed by combining the leakage power values for various conditions of $A1$ and $A2$ pins specified in the library. This computation is based upon the static probability values at the $A1$ and $A2$ pins. The computation is illustrated below.

```
Leakage power:
= 26.1 * Prob(!A1 !A2) +
    33.0 * Prob(!A1 A2) +
    27.0 * Prob(A1 !A2) +
```

---

[2] The static probability values at the output of combinational gates is shown in Fig. 4.8.

```
    82.7 * Prob(A1 A2)
  = 26.1 * (1 - 0.6) * (1 - 0.55)+
    33.0 * (1 - 0.6) * 0.55 +
    27.0 *0.6 * (1 - 0.55) +
    82.7 * 0.6 * 0.55

  = 46.539nW
```

4.2.1.2 Active Power Computation

The active power is computed based upon the 7.7 million transitions per second toggle rate at *ZN* and the 5 and 6 million transitions per second toggle rates on *A1* and *A2* respectively.

Internal Power

For the internal power due to switching activity on *ZN*, the appropriate path-dependent internal power table has to be used. In particular, the 7.7 million transitions per second toggle rate at *ZN* is mapped to path-specific (*A1->ZN*) or (*A2->ZN*) based upon the toggle rates of *A1* and *A2*. The distribution of the *ZN* toggles into path-specific toggles uses the same ratio as the toggle rates of inputs *A1* and *A2*.

A1->ZN toggle rate:
```
   = ZN toggle rate * A1 toggle rate /
     (A1 toggle rate + A2 toggle rate)
   = 7.7 * 5 / (5 + 6) million transitions/sec
   = 3.5 million transitions/sec
```

A2->ZN toggle rate:
```
   = ZN toggle rate * A2 toggle rate /
     (A1 toggle rate + A2 toggle rate)
   = 7.7 * 6 / (5 + 6) million transitions/sec
   = 4.2 million transitions/sec
```

A1 toggle rate not causing output transition:
```
   = A1 Toggle rate - A1->ZN toggle rate
   = (5 - 3.5) million transitions/sec
   = 1.5 million transitions/sec
```

A2 toggle rate not causing output transition:
```
   = A2 toggle rate - A2->ZN toggle rate
   = (6 - 4.2) million transitions/sec
   = 1.8 million transitions/sec
```

For *A1->ZN* toggle rate, we use *A1->ZN* internal power table and for *A2->ZN* toggle rate, we use *A2->ZN* internal power table. As described in Chap. 2, the internal power tables can be a nonlinear table defined in terms of input slew and output capacitance. However, for simplifying the explanation, the power values for this example are depicted as scalar values independent of input slew or the output capacitance.

The library description for the cell specifies internal power from each pin for the two scenarios:

1. When the input pin switching causes an output transition
2. When the input pin switching does not result in an output transition.

The latter corresponds to the condition "= !A2&ZN" for transition on input *A1* and condition "= !A1&ZN" for transition at input *A2*.

From the library description:

Internal power[3] **for transitions at *A1* which do not result in output pin transition:**
     = 0.004pJ (for rise transitions) and
       0.006pJ (for fall transitions).

Total internal power for transitions at *A1* which do not result in output pin transitions:
     = (1.5 million / 2) * 0.004
     + (1.5 million / 2) * 0.006
     = 7.5nW

In above, the transition rate is divided by 2 to obtain the rise transition rate and the fall transition rate. Again from the library description:

Internal power for transitions at *A1* resulting in output transition:
     = 0.043pJ (for output rise) and
       0.016pJ (for output fall).

Total internal power for transitions at *A1* resulting in output pin transition:
     = 3.5 million * (0.043 + 0.016) / 2
     = 103.25nW

   Total internal power due to transitions at *A1*:
     = 7.5 + 103.25
     = 110.75nW

Similar computation for input pin *A2* follows.

---

[3] As described in Chap. 2, the library power models actually represent the energy dissipated per transition.

Internal power for transition at A2 which do not result in output pin transition:
>   **= 0.006pJ (for rise transitions)** and
>   **0.008pJ (for fall transitions).**

Total internal power for transitions at A2 which do not result in output pin transitions:
>   = 1.8 million * (0.006 + 0.008) / 2
>   = **12.6nW**

Internal power for transitions at A2 resulting in output transition:
>   **= 0.036pJ (for output rise)** and
>   **0.021pJ (for output fall).**

Total internal power for transitions at A2 resulting in output pin transition:
>   = 4.2 million * (0.036 + 0.021) / 2
>   = **119.7nW**

Total internal power due to transitions at A2:
>   = 12.6 + 119.7
>   = **132.3nW**

Output Charging Power

Now we illustrate the computation of the output charging power. Assume that the power supply *Vdd* is 1.0 V and the output capacitance driven by *ZN* is 20fF.

>   **Toggle rate at pin ZN**
>   **= 7.7 million transitions/sec**

>   **Total output charging power:**
>   = 0.5 * C * Vdd * Vdd * Toggle rate
>   = 0.5 * 20fF * 1 * 1 * 7.7 million
>   = **77nW**

4.2.1.3   Total Power

The total power dissipation is the sum of the leakage power and active power. Using the values computed above:

>   **TOTAL POWER DISSIPATION in nand cell:**
>   = Leakage power + Internal power +
>   Output charging power
>   = 46.539 + (110.75 + 132.3) + 77
>   = **366.589nW**

Fig. 4.5 Power computation
using clock transition times
and pin activity information

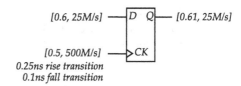

In each of the above cases, the *rise* and *fall* toggles are assumed to be equal. Thus in each computation, 50% of the toggles correspond to *rise* power models and 50% of the toggles correspond to the *fall* power models.

4.2.2   Power Computation for a Flip-Flop Cell

This section illustrates the power calculation of a D-type flip-flop cell using the switching activity information for various pins of the macro. The example illustrates the internal power calculation—the leakage and the output charging power components can be added similar to the case of the *nand* cell example in the previous subsection. The switching activity values at the pins of the flip-flop are shown in Fig. 4.5.

The flip-flop is clocked by a 250 MHz input clock with input transition times of 0.25 ns for rise and 0.1 ns for fall. An example fragment of the power specification for a D-type flip-flop cell is given below.

```
pin (CLK) {
  internal_power () {
    when : "(D&Q) | (!D&!Q)"; /* No transition on Q */
    rise_power (template_2x1) {
    index_1 ("0.1, 0.4"); /* Input transition */
      values ( /*          0.1              0.4 */ \
               "      0.050,      0.090");
    }
    fall_power (template_2x1) {
      index_1 ("0.1, 0.4");
      values ( \
        "0.070, 0.100");
    }
  }
  internal_power () {
    when : "(D&!Q) | (!D&Q)"; /* Has transition on Q */
    rise_power (scalar) {
     values ( "0" );
    }
```

```
    fall_power (template_2x1) {
      index_1 ("0.1, 0.4");
      values ( \
        "0.070, 0.110");
    }
  }
}
  pin (D) {
    direction: input;
    internal_power () { /* Input pin power */
    rise_power (scalar) {
      values ("0.026");}
    fall_power (scalar) {
      values ("0.011");}
    }
  }
  pin (Q) {
    direction: output;
    related_pin: CLK;
    internal_power () { /* When output switching */
    rise_power (scalar) {
      values ("0.09");}
    fall_power (scalar) {
      values ("0.11");}
    }
  }
```

The switching activity information for the signals at the *D*, *CLK*, and *Q* pins of the flip-flop is described as follows.

```
Static probability (pin D) = 0.6
Static probability (pin CLK) = 0.5
Toggle rate (pin D) = 25 million transitions/sec
Toggle rate (pin CLK) = 500 million transitions/sec
Static probability (pin Q) = 0.61
Toggle rate (pin Q) = 25 million transitions/sec
```

This corresponds to a flip-flop clocked with a 250 MHz clock with the input data and flip-flop output having 10% activity (that is, the flip-flop toggles in 10% of the clock cycles). The active power computation for the above scenario is described as follows.

```
Internal power due to transitions at input pin D
  = 25 million * (0.026 + 0.011) / 2
  = 0.4625µW
```

Fig. 4.6 Activity information at the pins of the single port memory macro

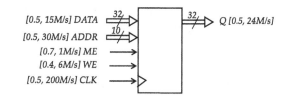

[0.5, 15M/s] DATA
[0.5, 30M/s] ADDR
[0.7, 1M/s] ME
[0.4, 6M/s] WE
[0.5, 200M/s] CLK

Q [0.5, 24M/s]

The internal power dissipation due to *CLK* pin transitions requires the breakdown of the *CLK* pin transitions into those which cause a transition at the output pin *Q* and the ones which do not create a transition at output pin *Q*. Based upon the activity at *CLK* and at *Q*, we can determine that amongst the toggle rate of 500 million transitions per second at the *CLK* pin, 25 million transitions per second (rise transitions) create a transition at the output pin *Q* and the remainder 475 million transitions per second (225 million *rise* transitions per second and 250 million *fall* transitions per second) do not cause a transition at the output pin *Q*.

Internal power due to output pin *Q*transitions:
```
= 25 million * (0.09 + 0.11)/2
= 2.5µW
```

Internal power due to *CLK* pin rise transitions:
```
= 25 million * 0.0 + 225 million * 0.07
= 15.75µW
```

Internal power due to *CLK* pin fall transitions:
```
= 250 million * 0.07
= 17.5µW
```

Total internal power:
```
= (0.4625 + 2.5 + 15.75 + 17.5)µW
= 36.2125µW
```

4.2.3   Power Computation for a Memory Macro

This section describes the power computation for an SRAM macro. We use the SRAM instance with the library as described in Sect. 3.1.1 to illustrate the power computation. The SRAM macro and the activity information for the signals at the pins of the SRAM macro are depicted in Fig. 4.6.

The activity values are:

```
CLK pin (100 MHz):
   200 million transitions/sec (for rise and fall)

Address pins:
```

30 million transitions/sec (for rise and fall)
Static probability: 0.5

Data pins:
15 million transitions/sec (for rise and fall)
Static probability: 0.5

Memory enable (*ME*):
1 million transitions/sec (for rise and fall)
Static probability: 0.7

Write enable (*WE*):
6 million transitions/sec (for rise and fall)
Static probability: 0.4

Output bus *Q* pins:
24 million transitions/sec (for rise and fall)

Internal power due to activity at one address pin:
= 30 million * (0.124 + 0.124) / 2
= **3.72μW**

Internal power due to activity at all 10 address pins:
= 3.72 * 10
= **37.2μW**

Internal power due to activity at one data input pin:
= 15 million * (0.153 + 0.153) / 2
= **2.295μW**

Internal power due to activity at 32 data input pins:
= 32 * 2.295
= **73.44μW**

Internal power due to activity at *ME* pin:
= 1 million * 0.048
= **0.048μW**

Internal power due to activity at *WE* pin:
= 6 million * 17.08
= **102.48μW**

For power due to clock pin, we need to compute the toggles:

WRITE (Rising *CLK* with both *ME* and *WE* high):
= CLK_rise_toggles * static_probability_of_ME *
static_probability_of_WE
= 100 million * 0.7 * 0.4
= **28 million transitions/sec**

READ (Rising *CLK* with *ME high*; *WE low*):
  = 100 million * 0.7 * 0.6
  = **42 million transitions/sec**

INACTIVE (Rising *CLK* with *ME* is *low*):
  = 100 million * 0.3
  = **30 million transitions/sec**

Based upon above, the clock power for each case is computed. Note that the falling clock transitions have negligible power dissipation and thus the library description for the memory macro in Sect. 3.1 shows zero power for falling transitions on the *CLK* pin. Thus, the power for various cases is computed as:

WRITE case:
  = 28 million * 42.4
  = **1187.2µW**

READ case:
  = 42 million * 43.9
  = **1843.8µW**

INACTIVE:
  = 30 million * 0.93
  = **27.9µW**

Total clock power:
  = 1187.2 + 1843.8 + 27.9
  = **3058.9µW**

For internal power due to output switching:

Internal power for each output pin switching:
  = 24 million * (0.022 + 0.022) / 2
  = **0.528µW**

Internal power for all 32 output pins:
  = 32 * 0.528
  = **16.896µW**

The output charging power corresponds to switching the output load capacitance at the output pins.

For each output pin, the output charging power:
  = **0.5 * C * Vdd * Vdd * Toggle_rate**

Assume each output drives a 20fF capacitance load and the power supply is 1.0 V.

Total output charging power for all 32 output pins:
  = 32 * 0.5 * 20fF * 1 * 1 * 24 million
  = **7.68µW**

Total active power in the memory macro:
```
= 37.2 + 73.44 + 0.048 + 102.48 + 3058.9 + 16.896
+ 7.68
```
= 3296.644µW

This illustrates that, in a practical scenario, the dynamic or active power of a memory macro is largely governed by the *read* and *write* power.

4.3 Specifying Activity at the Block or Chip Level

This section specifies various alternatives for specifying the switching activity information at the block or at the full-chip level.

4.3.1 Default Global Activity or Vectorless

This method is typically utilized for blocks in the initial design phase or where the designer does not have any detailed information. In this method, the designer provides an estimate of the activity ratio for all nets. Based upon the clock frequency and an estimated activity factor (for example 20% or 30%), the transition rate is obtained for all signals. The static probability can be specified or, in most cases, is set to a default value of 0.5. The static probability and the transition rate together constitute the *switching activity* information used for power analysis.

4.3.2 Propagating Activity from Inputs

This is typically the default method and it can be pessimistic in some cases. In this method, the clock nets toggle at the frequency specified in the SDC. When the clock hits a sequential element, the output of the sequential element gets a toggle rate depending upon the activity ratio specified for that clock. The flip-flops which get multiple clocks get the toggle rate of the fastest clock.

Nets with no clock phase[4] on them are set to zero activity, unless a default activity has been specified for all nets. Typically, a power analysis tool provides a way to annotate a net with a probability and a toggle rate.

When propagating the activity through combinational logic cells, the functionality of the cell is utilized to obtain switching activity at the cell output based upon the switching activity at the inputs of the cell. This can be accomplished by using heuristics such as cycle-based random simulation or it can be as simple as using the sum of the input transition rates as the transition rate for the output.

---

[4] Nets that have no transitions.

Fig. 4.7 Toggle rate at Q is at most one-half the toggle rate of *CLOCK*

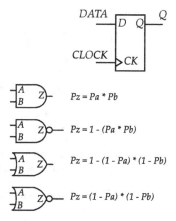

Fig. 4.8 Computing the static probability at a logic gate output

For a D-type flip-flop, as shown in Fig. 4.7, the transition rate at Q is at most half of that of clock *CLOCK*.

The static probability at the output of a combinational logic cell is computed based upon the static probability values at the input pins. As an example, Fig. 4.8 shows how the static probability is calculated at the output pins of 2-input logic cells. The static probability Pz at the output pin is a simple function of the static probability values, Pa and Pb, at the two input pins.

For combinational logic cells, the static probability values affect only the leakage power calculation. These values do not affect the internal and output capacitance charging (*swcap*) power numbers, which are governed by the transition rate or the switching activity.

In an RTL description, default global activity can be specified at the primary inputs and on outputs of all macros (as these macros do not change during synthesis). Activity values can then be propagated from these points onwards.

4.3.3   VCD

The VCD (Value Change Dump) is an ASCII output dump from logic simulation. The user can select the list of signals (and the time span) for the dump from a logic simulation tool.

The VCD output is a very accurate representation of the logic activity within the design. However, it is very time consuming to create a representative input vector set (*aka* test bench) for simulation. In general, the vector set used for functional verification is not representative for the purposes of power dissipation in the block. If the designer can create an input vector set that is representative in terms of power dissipation, the VCD output would be an accurate representation of the internal activity.

A sample VCD for a simulation fragment is depicted in Fig. 4.9. This fragment of the VCD shows only a few nodes from a design containing a large number of

$date	**$dumpvars**
Fri Sep 27 16:23:58 1996	1#
$end	0$
$version	b1 !
Verilog HDL Simulator 1.0	b10 "
$end	b101 +
$timescale	1(
100ps	0'
$end	1&
$scope module Test **$end**	1)
$var parameter 32 ! ON_DELAY	0*
$end	**$end**
$var parameter 32 " OFF_DELAY	#10
$end	0#
$var reg 1 # Clock **$end**	0)
$var reg 1 $ UpDn **$end**	#30
$var wire 1 % Cnt_Out [0] **$end**	1#
$var wire 1 & Cnt_Out [1] **$end**	1)
$var wire 1 ' Cnt_Out [2] **$end**	b100 +
$var wire 1 (Cnt_Out [3] **$end**	b101 +
$scope module C1 **$end**	#40
$var wire 1) Clk **$end**	0#
$var wire 1 * Up_Down **$end**	0)
$var reg 4 + Count [0:3] **$end**	#60
$var wire 1) Clk **$end**	1#
$var wire 1 * Up_Down **$end**	1)
$upscope $end	b100 +
$upscope $end	b101 +
$enddefinitions $end	#70
#0	0#
(continued next column)	. . .

Fig. 4.9 VCD example

nodes. The *$var* statements define mnemonic symbols for the signal names in the design whose values are being dumped. The *#<number>* specifies the timestamp at which the values are dumped. In between two such timestamps, the signal values are listed one at a time using their mnemonic symbol; these are of the form *<value><mnemonic symbol>*.

One drawback of the VCD format is that the output files for a complete design can be very large. To reduce the complexity of the VCD output, the VCD can be dumped between specific times or between specific vector sets. A more complete description of VCD can be found in [BHA06].

In terms of usage for power analysis, the VCD output is typically processed to produce a compact representation, such as *S*witching *A*ctivity *I*nterchange *F*ormat (SAIF), that captures the switching activity on a per net basis. The VCD to SAIF conversion can be specified on all or a group of nets (e.g. nets within a hierarchical block). The SAIF is used for detailed power analysis and the details of SAIF are described in the next section.

4.3.4 SAIF

SAIF is an ASCII format used to store the switching activity of a design. It is part of the IEEE Standard 1801 for the design and verification of low power integrated circuits.

Here is an example of a SAIF file fragment for a representative block.

```
(VERSION "1.0")
(DIVIDER / )
(TIMESCALE 1 ns)
(DURATION 8630.00)
(INSTANCE tb
  (INSTANCE dut
    (NET
      (clk
        (T0 4315) (T1 4315) (TX 0) (TC 6904) (IG 0)
      )
      (int6
        (T0 4457) (T1 4173) (TX 0) (TC 874) (IG 0)
      )
      (int8
        (T0 3978) (T1 4652) (TX 0) (TC 761) (IG 0)
      )
    )
  )
)
```

Most of the lines in a SAIF file are self-explanatory. Here are some definitions of the keywords in the SAIF file.

- $T0$: Total time duration when the net is at logic-0.
- $T1$: Total time duration when the net is at logic-1.
- TX: Total time duration when the net is at 'X'.
- TZ: Total time duration when the net is floating (no drivers).
- TB: Total time duration when the net is in a bus contention state (two or more drivers are driving it).
- TC: Toggle count or the total number of rise and fall transitions.
- TG: Number of transport glitches, or the number of 0-1-0 and 1-0-1 glitches (extra transitions at the output of a gate before it reaches its steady state).
- IG: Inertial glitches, or the number of 0-x-0 and 1-x-1 glitches (signal transitions that can be filtered out).
- IK: Inertial glitch derating factor.

The *timing attributes* are *T0*, *T1*, *TX*, *TZ* and *TB*. The *toggle attributes* are *TC*, *TG*, *IG* and *IK*.

A power analysis tool uses the timing and toggle attributes in SAIF to obtain the switching information (static probability and transition rate as described in Sect. 4.1) for each signal in the design. The static probability corresponds to the ratio of *T1* (time the net is at 1) to *DURATION* which is the total simulation time included in SAIF. Similarly, the transition rate is the ratio of *TC* (toggle count) to *DURATION* (the total simulation time included in SAIF).

4.3.4.1 State-Dependent and Path-Dependent Attributes

State-dependent timing attributes can be specified to indicate the condition under which the transitions occur. Here is an example.

```
(COND (!A * !B)       (T1 15)        (T0 7)
 COND (!A * B)        (T1 10)        (T0 12)
 COND_DEFAULT         (T1 5)         (T0 17))
```

The conditions determine a *priority encoded* specification of the timing attributes. The subsequent timing conditions are applicable only when the corresponding condition holds *and* all the previous conditions do not hold. In the above example, the COND_DEFAULT applies when none of the previous conditions are applicable. Thus, in the above example, the net is at 1 for 30 units of time and at 0 for 36 units of time. The toggle attributes can also be described in a state-dependent manner as described next.

State-dependent toggle attributes are specified as shown in an example below.

```
(COND WE (RISE) (TC 10)
 COND WE (FALL) (TC 9)
 COND RW (RISE) (TC 5)
 COND RW (FALL) (TC 6))
```

Similar to the state-dependent timing attributes, the above conditions determine a priority encoded specification of the toggle attributes. Of the total toggle count of 30, 15 are rise transitions, 10 occur when *WE* is 1, and 5 occur when *WE* is 0 and *RW* is 1.

The *path-dependent toggle attributes* can be specified as shown in an example below.

```
( IOPATH  CE (TC 20)
  IOPATH  ME RW (TC 15))
```

Of the 35 toggles, 20 are caused due to transitions on pin *CE*, and 15 are caused due to transitions on either *ME* or *RW*.

Various attributes in a SAIF file can be specified for the nets or the ports of the design.

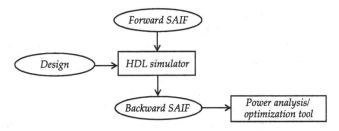

Fig. 4.10   Forward and backward SAIF

4.3.4.2   Backward or Forward SAIF Specification

The examples shown above correspond to the backward SAIF. A *backward SAIF* file is generated by an HDL simulator and it contains the switching activity that can be back-annotated into power analysis or other optimization tools.

A *forward SAIF* file contains directives on the format of the backward SAIF file. As shown in Fig. 4.10, the forward SAIF file is read by the HDL simulator, which then produces a backward SAIF file that contains data in conformance to the forward SAIF file. There are two kinds of forward SAIF files:

(a) *Library forward SAIF* file: Contains directives for generating state-dependent and path-dependent switching activity.
(b) *RTL forward SAIF* file: Contains directives for generating switching activity from the simulation of RTL descriptions.

The complete SAIF syntax is described in Appendix A. There is another format for providing activity information called the Toggle Count Format (TCF). This is a proprietary format, but has very similar information as in a SAIF file.

4.4   Power Analysis at Chip Level

4.4.1   Selecting the PVT Corner

Besides the switching activity, the power dissipation in a design varies significantly with the PVT (*P*rocess, *V*oltage or power supply, and *T*emperature) conditions used for analysis.

In digital designs, the power dissipation is highest at the following PVT conditions:

(a) *Process*: *Fast*—normally referred as FF (*FastN, FastP*) which refers to each MOS device (NMOS or PMOS) being at the fastest possible corner within the manufacturing process. At the FF process, both the *device-on* as well as the *-device-off* currents are high. The *on* currents map to the active currents and the *off* currents map to the leakage currents. Thus, both active and leakage power are high for the FF process condition.

(b) *Power supplies*: Maximum allowed for the design. In many cases, this may correspond to 10% above the nominal supply values. This refers to all the power supplies used for the design. In general, the power supply values are uncorrelated; however any correlation should be considered for analysis. The active power as well as leakage power increases with power supply.

(c) *Temperature*: Maximum junction temperature allowed for the design. In many cases, the maximum temperature allowed is 125°C. The leakage power has a superlinear (almost exponential) dependence on temperature. In most digital designs, the active power also increases with temperature though the increase is much smaller than that of leakage power.

Based upon the above, the power analysis using the above PVT corner will correspond to the worst (or maximum) power dissipation for the ASIC. A designer may also want to obtain the nominal power dissipation in which case, additional power analysis using nominal PVT libraries can be performed.

4.4.2   Power Analysis

4.4.2.1   Estimation in Very Early Design Phase

This refers to a simple spreadsheet analysis used for power estimation in the very early phase of the design. In the initial phase, the designer may only have high-level design information relating to design complexity. A power estimate can be obtained using the high-level design information such as gate count, technology, memory macro types and their activity, clock frequency (or frequencies), along with the flip-flop count. This along with the list of various IO interfaces can produce a reasonable power estimate for system planning purposes.

4.4.2.2   Analysis at Various Stages of Implementation

For power analysis, the design can be annotated with a wireload estimate or with actual RC parasitics specified through SPEF. The parasitic annotation, in conjunction with the switching activity specified from SAIF or VCD, forms the basis for detailed power analysis.

Quite often, the activity file does not capture the switching activity for all the nets in the design. In such cases, the power analysis is typically based upon performing zero-delay simulation to propagate the switching activity through all the non-annotated nets. A good activity file for performing this simulation should include activity of:

1. All black-boxes,
2. All inputs, and
3. All flip-flop outputs.

The activity description can be obtained from simulation on the RTL or the pre-layout netlist. The power analysis tool often requires the definition of clocks in the design. Note that nets, not pins, are annotated with switching activity.

Here are some typical values for default switching activity.

- Constant net: The toggle rate is 0 and the static probability is 0 or 1.
- Clock net: Follows directly from the clock definition.
- Buffer: The nets at the buffer input and output have the same switching activity.
- Inverter: The nets at the inverter input and output have the same toggle rate, but static probability value at the output is complement (of 1) of the value at the input pin.
- Flip-flop outputs: The toggle rate of Q is generally the same as that of input data. The toggle rates of Q and QN are identical, however the static probability of Q is complement of the value at QN.
- Black-box pins: These usually use a system-wide default activity and a static probability of 0.5.

4.5 Summary

This chapter described various methods for specifying the activity information for various nets in a design. Based upon the activity information for the nets, the leakage power and the active power are computed for all the elements in the design. Detailed power calculations for sample combinational and sequential cells and for memory macro are provided.

SAIF is a standard for specifying the activity information. Without the detailed activity information in the SAIF, default activity can be used to obtain the power information for the design.

Power analysis for a design can be performed at pre-layout phase, post-layout phase or at any intermediate stage of the design implementation. However, the more information about physical implementation is available, the more accurate the power analysis becomes.

Chapter 5
Design Intent for Power Management

This chapter describes the various features involved in specifying the design intent for a low power design and how these are specified for detailed implementation.

5.1 Power Management Requirements

There are several types of power management features used to specify the power design intent. Examples of these features are[1]:

1. Number of separate power domains in the design (which instances belong to which power domains)
2. Different voltage supplies for different portions of the design
3. Dynamically varying voltage for power supplies
4. Shutdown blocks
5. State retention
6. Power domain sequence for turn-on and turn-off
7. Power ground nets and ports
8. Cells to use to manage power domains
9. Power modes and mode transitions

5.2 Power Domains

A power domain normally refers to a part of design that operates at the same (or common) power supply. In a general design, different blocks can operate at different voltages to achieve the power objective. Level shifters are required to transfer data

---

[1] Most designs would use only a subset of these features.

R. Chadha and J. Bhasker, *An ASIC Low Power Primer: Analysis, Techniques and Specification*, DOI 10.1007/978-1-4614-4271-4_5, © Springer Science+Business Media New York 2013

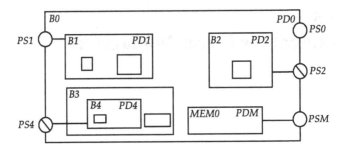

Fig. 5.1 Power domains and blocks

from one block to another block if the power supply voltage values of the two blocks are different. Similarly, isolation cells are required when transferring data from one block to another block if power to one of the blocks can potentially be shut down. This is to ensure that the receiving block gets a *valid*[2] signal even when the driving block has been turned *off*.

What is a power domain? It represents a grouping of blocks or logic with one common power supply. Each such group of blocks or logic is associated with one power domain.

Consider Fig. 5.1. Logic in block *B0* belongs to the power domain *PD0* which is connected to a power supply *PS0*, say 1.0 V. Block *B0* contains three other blocks and a memory *MEM0*. Block *B1* belongs to the power domain *PD1* which is connected to a power supply *PS1*, let's say it can be 0.9, 1.0 or 1.1 V. Block *B2* belongs to the power domain *PD2* which is connected to a power supply *PS2*, say 1 V, which can be turned off. Block *B3* belongs to the power domain *PD0*. However, *B3* contains another block *B4* which belongs to the power domain *PD4* which is connected to a power supply *PS4*, say 1.1 V, which can be turned off. The memory logic *MEM0* belongs to a power domain *PDM* which is connected to a power supply *PSM*, say 0.95 V.

As shown in the above example, a power domain may be shut down to save power. In such cases, isolation cells are needed on signals that cross boundaries between the potentially shut down domain and the *on* domain to ensure that there is no leakage problem.[3] The isolation cell ensures that a valid signal (*VSS* or *VDD*) is present at the input of the receiving block.

---

[2] A *valid* signal implies a logic-0 (VSS) or a logic-1 (VDD) for the receiving block, whereas an invalid signal may be a non-driven signal which can just be floating and/or be at an arbitrary intermediate voltage value.

[3] A CMOS cell with input values which are not "near" *VSS* or "near" *VDD* can result in increased power dissipation. This is because both the pull-up (NMOS) and the pull-down (PMOS) stages of the CMOS logic can be *on* simultaneously resulting in large crowbar current.

Table 5.1 Power state table

Operating modes and states	PS1	PS2	PS4
Mode1	0.9 V	OFF	1.1 V
Mode2	1.1 V	1.0 V	OFF
Mode3	1.0 V	1.0 V	1.1 V

Thus, a power domain is a logical partition with a common set of power characteristics. A power domain contains power nets, and has the same power down or power switching characteristics. A power domain can:

(a) Operate at different voltages depending on the state of the design, or
(b) Operate at a single voltage always, or
(c) Be completely shut down at some times (so that the power supply seen by the block is turned off), or
(d) Any combination of the above

For example, consider a design with three power domains, *PDa*, *PDb* and *PDc*. *PDa* operates at 1.0 V and can be shut down. *PDb* operates at 1.1, 1.0, or 0.9 V, while *PDc* is always-on and operates at only 0.9 V.

5.2.1 Power Domain States

A design can have multiple power domains. Each power domain can operate in multiple states, examples of which are listed next.

(a) Power domain can be switched off.
(b) Voltage can be reduced for a power domain when it is not required to operate at full speed.
(c) *Dynamic voltage scaling* (DVS): Voltage of power domain is adjusted dynamically based upon certain conditions. For example, the devices (i.e. chips) which are from a *faster* process lot can provide the target performance at a reduced voltage.
(d) *Dynamic frequency scaling* (DFS): Frequency of operation of logic within a power domain is adjusted dynamically based upon certain conditions.
(e) *Dynamic voltage and frequency scaling* (DVFS): Refers to combining the DVS and DFS techniques.

Table 5.1 shows an example of a power state table for the power domain example in Fig. 5.1. It shows the various operating modes of the design and for each operating mode, it defines the state of the power domain.

5.3 Special Cells for Power Management

To handle multi-Vdd low power designs, special cells are required during implementation. These cells are utilized for special function needs for multi-Vdd implementation where some blocks may optionally be powered down. These are:

1. Isolation cells
2. Level shifter cells
3. Power switch cells
4. Always-on cells
5. Retention cells
6. Standard cells with PG (power and ground) pins
7. Memories and other IP with PG pins

5.3.1 Isolation Cells

A net that connects an output of a cell in a shutdown power domain to a cell in an active power domain requires an isolation cell to prevent *crowbar*[4] current and spurious signal propagation. Isolation cells are typically placed on the outputs of the shutdown power domain. An isolation cell is used to prevent dead logic from driving active logic. For example, a *VDD*/2 voltage at the input of the *on* domain can result in a large power dissipation due to crowbar current. The goal of the isolation cell is to ensure that the input presented to the *on* domain has a *valid* voltage level.

Figure 5.2 shows isolation cells inserted on nets between the shutdown domain and the active power domain. An isolation cell can clamp its output to a value of logic-0 or a logic-1. The *and*-type isolation cells clamp their output to logic-0; when *power_ctrl* is logic-0, the output is forced to a logic-0. The *or*-type isolation cells

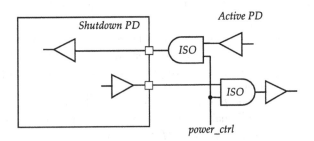

Fig. 5.2 Isolation cells required for a shutdown domain

---

[4] Crowbar current is the current flowing due to complementary PMOS and NMOS structures being *on* simultaneously.

also exist; those cells typically clamp their output to a logic-1. Note that the logic generating the *power_ctrl* signal should be placed in the active power domain.

Isolation cells may be required on the inputs to a shutdown domain in full-custom digital designs or where non-standard CMOS logic is used.

Here is an excerpt of a library file description[5] for an isolation cell.

```
cell(A2ISO_X4) {
  is_isolation_cell : true ;
  pg_pin(VDD) {
    voltage_name : VDD ;
    pg_type : primary_power ;
  }
  pg_pin(VSS) {
    voltage_name : VSS ;
    pg_type : primary_ground ;
  }
  pin(A) {
    direction : input ;
    input_voltage : default ;
    related_ground_pin : VSS ;
    related_power_pin : VDD ;
    isolation_cell_data_pin : true ;
  }
  pin(EN) {
    direction : input ;
    input_voltage : default ;
    related_ground_pin : VSS ;
    related_power_pin : VDD ;
    isolation_cell_enable_pin : true ;
  }
  pin(Y) {
    direction : output ;
    function : "(A&EN)" ;
    output_voltage : default ;
    related_ground_pin : VSS ;
    related_power_pin : VDD ;
    power_down_function : "!VDD + VSS" ;
  }
}
```

The cell-level attribute *is_isolation_cell* identifies that this is an isolation cell. The pin-level attribute *isolation_cell_enable_pin* identifies the control pin, and the *isolation_cell_data_pin* identifies the input data pin. The pin-level attribute *power_*

---

[5] Liberty file.

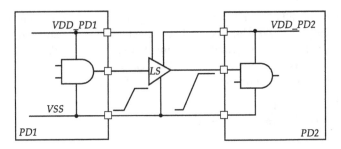

Fig. 5.3 Level shifter cell (LS) used when crossing power domain

down_function specifies the condition under which the output pin is switched off. This is an *and*-type isolation cell based on the function specified for the output pin *Y*. When *EN* is 0, the output of the isolation cell is clamped to a 0.

5.3.2   Level Shifters

A level shifter cell is used to change the voltage level of a signal. It can be used to raise the voltage level or to reduce the voltage level of a signal (see Fig. 5.3); a net that connects between two power domains (with different voltages) normally requires a level shifter cell in the middle.

There are three broad reasons for inserting level shifters between domains operating at different power supplies:

(a) *VDD_PD1 is less than VDD_PD2:* In this case, an input at logic-1 (voltage value *VDD_PD1*) can result in increased power dissipation due to crowbar current in the input CMOS cell in *PD2*. This is especially critical for scenarios where (*VDD_PD2* - *VDD_PD1*) can turn *on* the input pull-up devices. A level shifter ensures that the input presented to the receiving block (with voltage *VDD_PD2*) is valid, that is, the signal voltage transitions between *VSS* and *VDD_PD2*.

(b) *VDD_PD1 is greater than allowable gate-oxide reliability level for PD2 domain:* The designer should verify if the higher value *VDD_PD1* is within the acceptable range for the MOS gate-oxide level for the MOS transistors used in the *PD2* domain. If the voltage value for *VDD_PD1* is higher than the allowable voltage level for MOS devices used in the *PD2* domain, a level shifter must be used to prevent any oxide reliability issues for the input MOS devices in the *PD2* domain.

(c) *VDD_PD1 is greater than VDD_PD2 but still acceptable in terms of oxide reliability:* In such a case, a designer *may* still include a level shifter (a step-down level shifter) to ensure that the input buffer timing is correctly characterized.

Fig. 5.4 Potential level
shifter locations

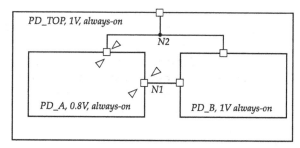

◁ : *potential for level shifter.*

Normally, the characterization of the standard cells would use the same input logic level as the power supply used for the cell. Thus, for cells operating at *VDD_PD2* supply, the timing characterization assumes that the inputs switch between *VSS* and *VDD_PD2*. The level shifter timing characterization ensures that proper input level (*VDD_PD1* for inputs) is used for timing characterization providing accurate timing analysis for the input paths.

Multi-voltage designs need level shifters to correct voltage level differences. A level shifter can reside logically and physically in either of the power domains.

Consider Fig. 5.4. It shows the possible locations of the level shifters for nets *N1* and *N2* that cross the power domains. No level shifters are required on net *N2* that crosses the *PD_TOP* and the *PD_B* domains as they are at the same voltage and are always-on. The same holds for net *N1*, that is, the level shifters are required only where the net crosses from *PD_A* into the *PD_TOP* domain. The level shifters can be placed inside the block or in the parent domain.

Here is a relevant excerpt from the library description of a level shifter.

```
voltage_map (COREVDD1,1.0)
voltage_map (COREVDD2,0.8)
. . . . .
cell (LVL_HL_X1) {
  is_level_shifter : true;
  level_shifter_type : HL;
  input_voltage_range (0.72, 1.1);
  output_voltage_range (0.72, 1.1);
  pg_pin (VDD) {
    pg_type : primary_power;
    voltage_name : COREVDD2;
    std_cell_main_rail : true;
  }
  pg_pin (VSS) {
    pg_type : primary_ground;
    voltage_name : COREGND1;
```

```
    }
    pin(I) {
      direction : input;
      input_signal_level : COREVDD1;
      level_shifter_data_pin : true;
      related_ground_pin : VSS;
      related_power_pin : VDD;
    }
    pin(Z) {
      direction : output;
      output_signal_level : COREVDD2;
      power_down_function : "!VDD + VSS";
      function : "I";
      related_ground_pin : VSS;
      related_power_pin : VDD;
    }
  }
```

The above is an example of a high-to-low level shifter. There are four cell-level attributes:

1. is_level_shifter
2. level_shifter_type
3. input_voltage_range
4. output_voltage_range

The cell attribute *is_level_shifter* identifies the cell to be a level shifter cell. The *level_shifter_type* attribute can take values *LH* (low to high), *HL* (high to low) or *HL_LH* (to indicate high to low and low to high). The *input_voltage_range* and *output_voltage_range* attributes provide the valid voltage ranges for input and output.

A level shifter can also have the following pin-level attributes:

1. std_cell_main_rail
2. level_shifter_data_pin
3. input_voltage_range
4. output_voltage_range
5. input_signal_level
6. power_down_function

The *std_cell_main_rail* attribute defines the power pin which is the main rail in the cell. This is used to determine at which side of the voltage boundary the level shifter is placed. The *level_shifter_data_pin* attribute specifies the input data pin of the cell. The *input_voltage_range* and *output_voltage_range* attributes can be specified for the pin as well. The *input_signal_level* attribute defines the overdrive level shifter cells. The *power_down_function* attribute specifies the boolean condition under which the output pin is switched off.

The level shifter type illustrated in the example is also referred to as a *buffer-type* level shifter.

Fig. 5.5 An enable level
shifter within the top-level
power domain

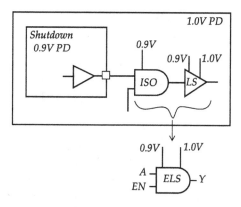

5.3.3   Enable Level Shifters

An enable level shifter combines the functionality of an isolation cell and a level shifter in one unified cell. Such a cell is used to provide isolation as well as level shifting when placed in the *on* domain (see Fig. 5.5). These cells are required for designs that have multi-voltage and shutdown regions.

Here is a relevant excerpt from the library description of an enable level shifter.

```
cell(A2LVLD_X1) {
  is_level_shifter : true ;
  level_shifter_type : HL ;
  input_voltage_range(0.72, 1.1);
  output_voltage_range(0.72, 1.1);
  pg_pin(VDD) {
    voltage_name : VDD ;
    pg_type : primary_power ;
    std_cell_main_rail : true ;
  }
  pg_pin(VSS) {
    voltage_name : VSS ;
    pg_type : primary_ground ;
  }
  pin(A) {
    direction : input ;
    input_signal_level : VDDI ;
    input_voltage : vddin ;
    related_ground_pin : VSS ;
    level_shifter_data_pin : true ;
  }
```

```
pin(EN) {
  direction : input ;
  input_voltage : ls_enable ;
  related_ground_pin : VSS ;
  related_power_pin : VDD ;
  level_shifter_enable_pin : true ;
}
pin(Y) {
  direction : output ;
  function : "(A&EN)" ;
  output_voltage : default ;
  related_ground_pin : VSS ;
  related_power_pin : VDD ;
  power_down_function : "!VDD + VSS" ;
}
}
```

All the attributes of a buffer-type level shifter are also applicable for enable level
shifters. The additional attribute of this cell is the *level_shifter_enable_pin*. This pin
attribute identifies the enable pin. When *EN* is 0, the output is clamped to a 0. The
enable level shifters that clamp to an output value of logic-1 also exist; these are
typically based upon *or*-logic.

5.3.4   Power Switches

A power switch cell provides the capability to shut off the power to a domain of
logic. A typical switch cell is shown in Fig. 5.6. *VDDG* is the primary power and

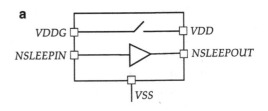

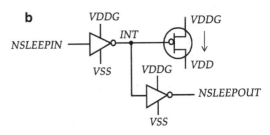

Fig. 5.6 A header type
switch cell

VDD is the switched power (also sometimes called virtual power). The *NSLEEPIN* signal controls the power switch. A delayed acknowledge signal *NSLEEPOUT* can be provided (but is optional). In a coarse-grain[6] power switch implementation, normally multiple power switches are used in each block that is to be switched off.

A power switch cell that is typically used is an HVt cell. In general, the switch cells are applied on a block level. A switch cell can be applied to *VDD*, called *header switch cells,* and/or can be applied to *VSS*, called *footer switch cells.* Here is an example excerpt from a library description of a header switch cell.

```
cell(HEADBUF_X16) {
   dont_touch : true ;
   dont_use : true ;
   switch_cell_type : coarse_grain ;
   leakage_power() {
      related_pg_pin : "VDDG" ;
      value : "0.0245051892" ;
   }
   /* IV curve information */
   dc_current (ivt125x25) {
      related_switch_pin : INT ;
      related_pg_pin : VDDG ;
      related_internal_pg_pin : VDD ;
      index_1("0, 0.0081, 0.0162, 0.0243, 0.0324, \
         ...0.7776, 0.7857, 0.7938, 0.8019, 0.81");
      index_2("0, 0.162, 0.324, 0.486, 0.648, 0.6561, \
         ...0.7776, 0.7857, 0.7938, 0.8019, 0.81");
      values("2.71498, 2.43708, 2.135, 1.77218",\
         ...
         " 1.39179e-05, 9.75069e-06, 5.11841e-06, 0");
   }
   pg_pin(VDD) {
      voltage_name : VDD ;
      pg_type : internal_power ;
      direction : output ;
      switch_function : "!NSLEEPIN" ;
      pg_function : "VDDG" ;
   }
   pg_pin(VDDG) {
      voltage_name : VDDG ;
      pg_type : primary_power ;
```

---

[6] Coarse-grain and fine-grain distribution of switches is described in Sect. 6.7. This section also describes *fine-grain* power gating where the switch can be applied to turn off individual cells.

```
  }
  pg_pin(VSS) {
    voltage_name : VSS ;
    pg_type : primary_ground ;
  }
  pin(INT) {
    direction : internal ;
    timing () {
    related_pin : "NSLEEPIN" ;
    timing_sense : negative_unate ;
    timing_type : combinational ;
    ....
    }
  }

  pin(NSLEEPIN) {
    direction : input ;
    input_voltage : header ;
    related_ground_pin : VSS ;
    related_power_pin : VDDG ;
    switch_pin : true ;
    always_on : true ;
  }
  user_function_class : HEAD ;
  }
```

The cell in Fig. 5.6 is a *header* type switch cell since the switch is placed on the *VDD* supply. When the sleep signal is active (*NSLEEPIN* is at logic-0 state) the switch is open and the *VDD* output is no longer logically connected to the primary power supply *VDDG*. Note that the standard cells have their power rails connected to *VDD* and *VSS* pins; thus the header type switch cells have the switched supply as *VDD* and the *true* or *always-on* supply as *VDDG*.

The switch cells are marked as *don't-use* so that synthesis tools would ignore (not use) these cells during synthesis. The switch cells have the *dont_touch* attribute, so that if any switch cells are included in a design, these cells are not modified (optimized away) by the design tools. The *switch_cell_type* attribute identifies this to be a coarse-grain power switch cell. This attribute cannot take any other value. The *dc_current* attribute provides the current flowing through the switch. This information is normally provided as a two-dimensional table in terms of the voltage value on the control signal (*INT*) and the voltage value on the switch power pin (*VDD*). The *dc_current* information can be used to compute the voltage drop through the switch based upon the power supply current drawn when the switch is *on*. Similarly, the *dc_current* attribute can be used to compute the leakage current through the switch when it is *off*. The *related_switch_pin* specifies the internal control signal controlling the MOS switch. The *related_pg_pin* specifies the true power

Fig. 5.7 A footer type switch cell

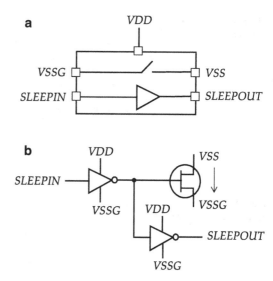

supply as *VDDG* (always-on power supply) and the *related_internal_pg_pin* attribute specifies the switched power supply as *VDD* (or virtual power supply).

The *switch_function* pin-level attribute specifies the condition under which the switch is turned off. The *pg_function* attribute specifies the functionality of the pin, and the *switch_pin* attribute identifies the control pin of the switch cell. The *NSLEEPIN* pin is marked with an attribute of *always_on* indicating that only always-on logic can drive this pin.

An alternate implementation of the switch cell may place the switch on the *VSS* supply; these cells are called *footer* type switch cells (see Fig. 5.7). When footer type switch cells are used, the power rails in the standard cells are still connected to *VDD* and *VSS* pins. In the footer cell, when the sleep signal is active (*SLEEPIN* is at logic-1 state), the switch is open and the *VSS* output is no longer logically connected to the ground supply. This is because the footer type switch cells have the switched ground supply as *VSS* and the *true* or *always-on* ground supply is *VSSG*.

The examples above are for switch cells that are single input header or footer cells, since there is only one control input—*NSLEEPIN* for header type or *SLEEPIN* for footer type switch cell. The optional output *NSLEEPOUT* (or *SLEEPOUT*) is the buffered output which can be used for daisy-chaining the switch cells. Other switch cells can have two control inputs (*NSLEEPIN1*, *NSLEEPIN2* or *SLEEPIN1*, *SLEEPIN2*) and similarly two control outputs. Figure 5.8 shows such an example. This is also referred to as a mother-daughter configuration of a switch cell. The *NSLEEPIN1* is made active first; this causes the daughter switch (weaker switch with larger *on*-resistance) to turn *on*. Subsequently, the *NSLEEPIN2* is turned *on*, causing the stronger mother switch (with lower *on*-resistance) to become active. The reverse typically occurs during shutdown. The advantage of this configuration is that the amount of in-rush[7] current can be controlled to two phases.

---

[7] Rush currents are described in detail in Sect. 7.6.

Fig. 5.8 Mother-daughter
switch cell configuration

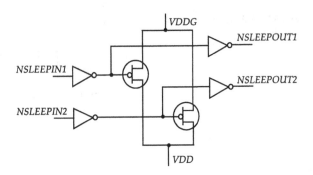

Fig. 5.9 Header type
always-on non-inverting cell

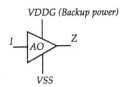

5.3.5   Always-on Cells

Control signals for power management cells need to be *live* even when the logic
around gets shut down. Optimizing and synthesizing such control signals with special
cells having secondary (or always-on) power is called *always-on synthesis*. An
always-on cell is a special cell that gets its power from the active domain, but is placed
in the shutdown power domain. The always-on cells are designed so that they can be
placed in the shutdown domain. Unlike the standard cells powered from *VDD* and *VSS*
rails which may be powered off, the always-on cells are powered from the active
power domain (with the *true* power supply *VDDG/VSS* or *VDD/VSSG*). Just like the
switch cells (described in the previous sub-section) can be *header* type or *footer* type,
the always-on cells need to be consistent with the switch cells used for the power
domain. Figure 5.9 shows an example of a *header* type always-on buffer cell.

An example of a *footer* type always-on inverter cell is depicted in Fig. 5.10.
Here is an excerpt of a library file for a *header* type always-on buffer cell.

```
cell (AO_X1) {
  always_on : true;
  pg_pin (VDD)⁸ {
    pg_type : primary_power;
    voltage_name : COREVDD1;
  }
  pg_pin (VDDG) {
```

---

[8] The *VDD* pin, even though it exists as a pin on the cell, does not typically connect to anything
within the cell.

Fig. 5.10 Footer type
always-on inverting cell

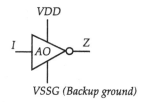

VSSG *(Backup ground)*

```
    pg_type : backup_power;
    voltage_name : COREVDD2;
}
pg_pin (VSS) {
    pg_type : primary_ground;
    voltage_name : COREGND1;
}
pin(I) {
    direction : input;
    related_ground_pin : VSS;
    related_power_pin : VDDG;
}
pin(Z) {
    direction : output;
    power_down_function : "!VDDG + VSS";
    function : "I";
    related_power_pin : VDDG;
    related_ground_pin : VSS;
}
}
```

A library may provide *footer* type always-on buffer cells as well. Similarly, a library could provide *header* or *footer* type—always-on inverting cells.

Always-on cells are used in shutdown power domains where some cells are needed to be always in an active state. Examples of these include enable pins, control to power switches and retention cells. Figure 5.11 shows an example where *header* type always-on cells are used to buffer the control signals to a retention flip-flop in a shutdown power domain.

Figure 5.12 shows another example where a signal needs to cross through a shutdown power domain to get to the other side. The always-on cells are needed to ensure that electrical design rules such as the load capacitance and slew rates are maintained and that the net continues to be active as it crosses the shutdown domain.

During the physical design implementation, certain category of cells are automatically treated as requiring always-on cells. Examples of these cells with automatic inferencing for always-on cells are the enable pins of enable level shifters and isolation cells, save and restore pins of retention cells, and the control pins of switch cells. For each such pin, cells and nets on its path are marked as always-on by tracing backwards until an always-on region is reached.

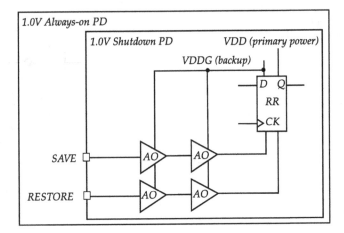

Fig. 5.11 Always-on cells to buffer the control signals in shutdown domain

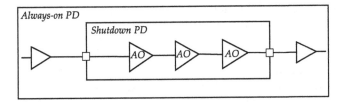

Fig. 5.12 Always-on cells needed when a net has to cross a shutdown domain

5.3.6   Retention Cells

Retention cells can retain their internal state even when the primary power supply is turned off. Retention cells are sequential cells and are of two types: a retention flip-flop and a retention latch. A retention cell is comprised of a regular flip-flop (or a latch) with an additional *save*-latch that holds the state when the primary power is shut down and can restore the state when the primary power is restored. Figure 5.13 shows an example of an implementation of a retention flip-flop. The save-latch normally uses HVt transistors so that the leakage during standby mode is low. In addition, the save-latch is powered with a backup power supply. During active mode, the flip-flop in the retention cell operates just like a regular flip-flop. In sleep mode, the Q data is transferred to the save-latch and the primary power supply to the flip-flop is turned off. This helps to save the flip-flop power in standby mode. When the restore signal arrives, the data in the save-latch is transferred back to the flip-flop.

Retention cells are used if the state of some shutdown logic needs to be preserved. Retention cells are useful for designs that want to recover their state after coming back from shutdown. A retention cell has both a primary power and a backup power supply (which is kept *on*).

Fig. 5.13 An implementation of a retention flip-flop

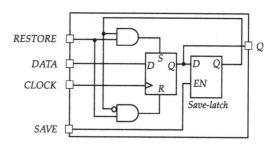

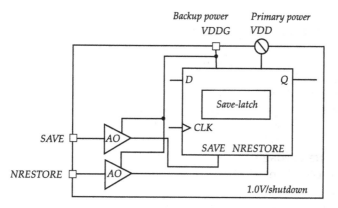

Fig. 5.14 Retention flip-flop in a shutdown domain

Figure 5.14 shows an example of a retention flip-flop placed in a shutdown domain with a main supply of 1.0 V. The retention flip-flop has a *VDDG* power pin that is used to connect the backup power which stays *on* even when the primary power supply *VDD* is turned *off*. Note that the backup power supply voltage can be reduced to save power during the retention mode. The backup power may operate at lower than the nominal supply—can be reduced, for example, to 0.6 V. *VDD* is the primary 1.0 V power pin which is turned off during shutdown. Note that to get to the *SAVE* and *NRESTORE* pins of this retention flip-flop, always-on cells (marked *AO*) have to be used; such always-on cells are connected to *VDDG* and are not turned off. The *SAVE* control signal would save the data in the retention latch and the *NRESTORE* control signal would restore the data from the retention latch.

Here is an excerpt from a library file of a retention flip-flop.

```
cell(DRFFQX0) {
  pg_pin(VDD) {
    voltage_name : VDD ;
    pg_type : primary_power ;
  }
  pg_pin(VDDG) {
```

```
      voltage_name : VDDG ;
      pg_type : backup_power ;
   }
   pg_pin(VSS) {
      voltage_name : VSS ;
      pg_type : primary_ground ;
   }
   ff(IQ,IQN) {
      clocked_on : "CK" ;
      next_state : "(D) (NRESTORE !SAVE)" ;
   }
   retention_cell : DRFF ;
   pin(Q) {
      direction : output ;
      function : "IQ" ;
      related_ground_pin : VSS ;
      related_power_pin : VDD ;
      power_down_function : "!VDD + !VDDG + VSS" ;
   }
   pin(NRESTORE) {
      direction : input ;
      related_ground_pin : VSS ;
      related_power_pin : VDDG
      retention_pin (restore, "1");
      always_on : true ;
   }
   pin(SAVE) {
      direction : input ;
      related_ground_pin : VSS ;
      related_power_pin : VDDG
      retention_pin (save, "0");
      always_on : true ;
   }
}
```

The *DRFFQX0* is a retention flip-flop that has the ability to retain its state when the primary power *VDD* is switched off. Power for the retention cell in that mode comes from the backup power *VDDG*. When the *SAVE* input is a 0, the flip-flop operates in normal mode. When *SAVE* is a 1, the state of the flip-flop gets saved at the active clock edge into the save-latch. When the *NRESTORE* is 0, the saved state is restored on the flip-flop output *Q*. Typically, it is illegal for both *SAVE* and *NRESTORE* to be active at the same time (*SAVE* to be a 1 and *NRESTORE* to be a 0).

The *retention_cell* cell-level attribute is used to specify the type of retention cell. The pin-level attribute *retention_pin* is used to identify the retention pin and to specify whether it is a *save*, *restore* or a *save_restore* pin. When the attribute

Fig. 5.15 Clock gate cell

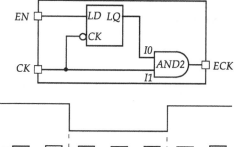

Fig. 5.16 Clock gater waveforms

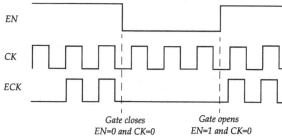

retention_pin uses the value of *save_restore*, then there is only one pin in the retention cell that serves the function of both save and restore.

In a retention latch, one difference is on the condition for the latch cell attribute *data_in*.

```
// In a normal latch:
data_in: D;

// In a retention latch:
data_in: D & (SAVE & RESTORE);
```

Retention cells are also called SRPG (State Retention Power Gating) cells.

5.3.7 Clock Gate Cells

So what is a clock gate? In today's technology, it is a standard cell that allows a clean start and stop of a clock. It basically consists of a latch and an *and*-gate as shown in Fig. 5.15.

The latch prevents any glitches on the enable pin *EN* from propagating to the output of the clock gater. When *CK* is a 0, the value of *EN* propagates to the output of the latch, but is blocked at the *and*-gate due to *I1* being 0. When *CK* becomes a 1, the value of *EN* is captured into the latch. If *EN* is a 1, the *CK* now propagates through the *and*-gate. If *EN* is a 0, then *CK* does not pass through the *and*-gate. Figure 5.16 shows the waveforms for the clock gater.

The above clock gater handles clocks that are connected to rising edge-triggered flip-flops (value is latched on falling edge so rising edge can propagate through). A similar clock gate cell exists that can handle falling edge-triggered flip-flops. Figure 5.17 shows an example of such a clock gate. In this case, the rising edge of *CKN* captures the value of *EN* so that the negative edge propagates through cleanly.

Fig. 5.17 Negative-edge
clock gate cell

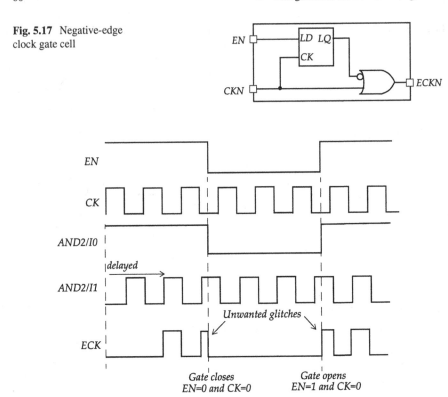

Fig. 5.18 Unwanted glitch at output of gater

In older technologies, integrated clock gating (ICG) cells were not available and
a clock gating cell was constructed using discrete logic gates. The disadvantage of
constructing such a cell was that it required tight constraints on how the logic was
placed on a chip; both the latch and the gate needed to be physically close to each
other. The real strict requirement was for the latch output to change only during the
quiet period of the clock at the inputs of the *and*-gate (or *or*-gate). Consider the
positive-edge clock gate cell shown in Fig. 5.15. Figure 5.18 shows the waveforms
if the delay of the clock to the *and*-gate is large, which causes an unwanted glitch to
occur at the output of the gater.

In newer technologies, an integrated clock gating cell is provided in the cell
library. The advantages of using an integrated clock gater are:

1. There is no clock skew between the latch and the *and*-gate.
2. Timing analysis and clock tree synthesis can handle the clock gates.
3. Setup and hold are modeled in the library file for the clock gating cell.

All of the above make it straightforward to use.

The disadvantages of using a discrete clock gater are:

1. Physical design has to ensure a minimum skew between the latch and the
   and-gate.

Fig. 5.19 Clock gate cell
with precontrol scan enable

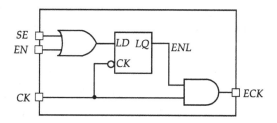

2. The latch clock pin needs to be handled specially during clock tree synthesis (it is not a balance point).
3. One must specify setup and hold check for the clock gater explicitly.
4. It adds complexity to the flow.

Here is an excerpt of a library file for an integrated clock gating cell shown in Fig. 5.19.

```
cell(CKICG_X4) {
  clock_gating_integrated_cell :
    "latch_posedge_precontrol" ;
  statetable("CK E SE", ENL) {
    table :     " L L L : - : L, \
                L L H : - : H, \
                L H L : - : H, \
                L H H : - : H, \
                H - - : - : N" ;
  }
  pin(CK) {
    clock : true ;
    clock_gate_clock_pin : true ;
    direction : input ;
  }
  pin(E) {
    clock_gate_enable_pin : true ;
    direction : input ;
  }
  pin(ECK) {
    clock_gate_out_pin : true ;
    direction : output ;
    state_function : "(CK&ENL)" ;
  }
  pin(ENL) {
    direction : internal ;
    internal_node : ENL ;
    inverted_output : false ;
  }
```

Fig. 5.20 Latch-free
and-clock gate cell

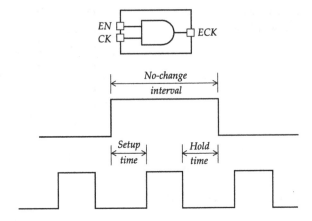

```
pin(SE) {
  clock_gate_test_pin : true ;
  direction : input ;
}
}
```

The attribute in the library file that identifies that it is a clock gating cell is *is_clock_gating_cell* or *clock_gating_integrated_cell*. The value of *latch_posedge_precontrol* in above library description specifies that this clock gating cell is for a positive-edge clock and that the test signal *SE* is connected before the latch (as opposed to after the latch, in which case it is called *postcontrol*). The internal pin *ENL* represents the state of the latch.

In a cell, the value of the *clock_gating_integrated_cell* attribute could also be *non_posedge* or *non_negedge*. These values signify a *latch-free clock cell* (same as Figs. 5.15 and 5.17 but without the latches). In such a clock cell, strong setup and hold requirements are imposed to ensure that the clock pulses are gated properly to ensure no glitch at the output. Figure 5.20 shows an example of such a latch-free *and*-clock cell (*clock_gating_integrated_cell* value is *non_posedge*).

No-change timing arcs are defined on the enable pin. The setup requirement on *EN* is defined with respect to the rising edge of clock *CK*, while the hold requirement is specified with respect to the falling edge of *CK*. The advantage of a latch-free clock gating cell is the power savings achieved from not having a latch.

5.3.8   Standard Cells

Standard cell libraries should be *pg-pin* ready, that is, the library description should include information on the *VDD* and *VSS* pins. Here is an excerpt of a library file for a standard cell.

```
library . . . {
  . . .
  voltage_map (VDD, 1.0);
  voltage_map (VSS, 0.0);
  default_operating_conditions: name_of_oc;
  . . .
  cell(AND2_X0) {
    leakage_power() {
      related_pg_pin : "VDD" ;
      when : "!A&!B" ;
      value : "0.005" ;
    }
    pg_pin(VDD) {
      voltage_name : VDD ;
      pg_type : primary_power ;
    }
    pg_pin(VSS) {
      voltage_name : VSS ;
      pg_type : primary_ground ;
    }
    pin(A) {
      input_voltage : default ;
      related_ground_pin : VSS ;
      related_power_pin : VDD ;
    }
    pin(B) {
      input_voltage : default ;
      related_ground_pin : VSS ;
      related_power_pin : VDD ;
    }
    pin(Y) {
      output_voltage : default ;
      related_ground_pin : VSS ;
      related_power_pin : VDD ;
      power_down_function : "!VDD + VSS" ;
      internal_power () {
        related_pg_pin: VDD;
        . . .
      }
    }
  } /* cell */
} /* library */
```

The library level attribute *voltage_map* defines the mapping between voltage names and their voltage values. The library attribute *default_operating_conditions* identifies

Fig. 5.21 Power-friendly memory

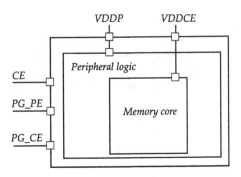

the PVT[9] conditions used to characterize this library. The *pg_pin* group cell-level attribute is used to describe the power and ground pins of the cell. The *voltage_name* attribute specifies the associated voltage name of the power and ground pin defined earlier with the *voltage_map* attribute. The *pg_type* attribute specifies the type of power and ground. It can have any one of the values: *primary_power*, *primary_ ground*, *backup_power*, *backup_ground*, *internal_power*, *internal_ground*; the *primary** are the *true* power and ground, the *backup** are the power and ground associated with retention or always-on cells for secondary/backup power, and *internal** are the power and ground associated with the output of switched cells.

The pin-level attributes *related_power_pin* and *related_ground_pin* are used to associate a predefined power and ground with the corresponding signal. The attribute *power_down_function* specifies the boolean condition under which the output pin is switched off. The *related_pg_pin* attribute (for leakage and internal power) associates the power data to a specific power pin.

The requirement to be *pg-pin* ready should be satisfied for all library files that are used in a multi-voltage design.

5.3.9 Dual Rail Memories

A memory that is power-friendly can optionally include a power gate to turn off the main power and/or have separate power supplies for the memory core and for the memory peripheral logic (see Fig. 5.21).

When the power gate enable *PG_PE* is *on*, the power to the peripheral logic is shut off. When the retention mode enable *PG_CE* is *on* and the *PG_PE* is *off*, a substrate bias is applied to the memory core array to reduce leakage. In this way, the memory contents are retained. In power down mode, both the peripheral logic and

---

[9] Process, Voltage, Temperature.

Table 5.2 Power mode table

Modes	CE	PG_PE	PG_CE
Standby mode	0	0	0
Power mode transition	0	1	0
Data retention	0	0	1
Power down	0	1	1
Normal operation	1	0	0

the memory core array are shut down and the memory contents are lost. Such a memory could have the power modes shown in Table 5.2.

A power sequence is typically required to take a memory from active mode to retention mode and back to active mode. For example, it may require going from active mode to standby mode to transition mode to retention mode. To get back to active mode, it may require going from retention mode to transition mode to standby mode to active mode.

5.4 Summary

This chapter provided a broad overview of different types of cells required for the specific requirements of power management. Specifically, it described scenarios when different types of cells are needed. For gating off clocks, clock gating cells are required.

In multi-voltage designs, level-shifters are required.

For a design that has a shutdown domain with no state retention, isolation cells and power switches are required.

In a design with multi-voltage domains and shutdown domains with no state retention, level shifters, isolation cells and power switches are required.

In a design with multi-voltage domains including shutdown domains with state retention, level shifters, isolation cells, power switches, retention registers and always-on cells are required.

Chapter 6
Architectural Techniques for Low Power

This chapter describes architectural techniques for achieving a low power design. There are numerous algorithms and techniques available to a designer and this chapter provides only a sampling of these techniques. There are also many techniques being evolved with some techniques targeted for specific design styles. This chapter should be viewed as a sampling of some the current design practices.

6.1 Overall Objectives

For the purposes of optimization of a design, it is important to understand the system related aspects and the overall objective and constraints related to power. A question the designer may consider is whether the goal is to minimize power or to minimize the total energy required to complete a computation. Alternately, is the goal to achieve highest performance within specific power constraints? If minimizing power, is the goal to minimize the average power or to minimize the peak power. Understanding the objective and the trade-offs enables the designer to adopt the right approach for implementation.

1. *Reduce power or energy.* As an example, consider two implementations (A and B) for a given data computation. Implementation A requires 20 clock cycles with 100 mW power dissipation whereas implementation B requires only 5 clock cycles with 200 mW power dissipation. While implementation A has a lower power requirement than B, the designer can architect a solution where the design is shut off after the computation is completed. In such a scenario, the B implementation has a lower energy requirement (200 mW power dissipation for 5 clock cycles vs. 100 mW power dissipation for 20 clock cycles). This shows that while implementation B requires 2× power when compared to A, the energy requirement of implementation B is only one-half or 50% of that of implementation A (Fig. 6.1). Either implementation may be chosen depending upon whether the

R. Chadha and J. Bhasker, *An ASIC Low Power Primer: Analysis,*
Techniques and Specification, DOI 10.1007/978-1-4614-4271-4_6,
© Springer Science+Business Media New York 2013

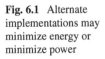

Fig. 6.1 Alternate implementations may minimize energy or minimize power

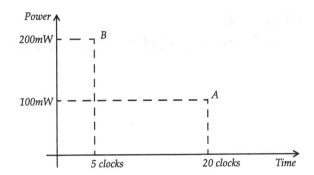

overall goal is to minimize power (choose *A*) or to minimize energy (choose *B*). Minimizing energy is very important for battery operated applications where battery life is a critical factor. Another factor critical in this selection is the allowed time to complete a computation—can the system wait 20 clock cycles for the lower power implementation?

2. *Reduce peak power or average power.* The requirement of minimizing the average power is essentially the same as minimizing the total energy required. The requirement for reducing peak power is driven by:

    (a) *Package/system/IP thermal considerations.* Operating temperature of the device should not exceed the allowed limits. This check should include all macros including IP, since the operating temperature for some IP macros may be limited.

    (b) *Limiting the IR drop.* Ensure that the operating voltage stays within the allowed limits.

3. *Reduce functional mode power or standby power.* If the operation of the device involves long durations where it is in standby mode, the power during standby mode dominates the usage. In such cases, minimizing standby power can be the critical objective for implementation. However, if the operation is nearly continuous in the functional mode, such as in a device operating from mains, the functional mode power can be the objective for optimization.

6.1.1 Parameters Affecting Power Dissipation

To understand how to reduce power (or energy), we need to first understand the parameters that control the amount of power dissipated in an ASIC.

Dynamic power is a function of net capacitance, supply voltage and toggle frequency. Thus, to reduce dynamic power, we need techniques that can reduce net capacitance, allow the design to operate at lower voltage, and allow operation of the circuit at lower frequencies. These techniques try to achieve one or a combination of these objectives to lower the dynamic power while still providing acceptable performance. Some of these techniques involve logic restructuring, sizing, reduced-VDD, multiple power supplies, clock gating, dynamic voltage frequency scaling (DVFS), area reduction and gate-level logic optimization.

Static power is a function of the supply voltage, the cells and macros used in the design, specifically on the strength of cells, number of cells, and whether SVt, LVt or HVt cells are used. Thus, to reduce static power, one can reduce the power supply voltage, reduce the number of cells in the design and/or use higher Vt cells in the design. Some of the techniques that achieve this are multi-Vt, multi-Vdd, and power gating.

As described above, a design may have an objective to minimize total power or total energy dissipated depending on the application. For example, a battery-run device may require both dynamic and static power to be minimized. However, when trying to optimize area, speed and power of a design, there is a trade-off between different objectives. Optimizing only for area may not achieve the speed objective, optimizing only for power also may not achieve the speed objective, optimizing for speed may not achieve the target power requirement, and so on.

6.2   Variable Frequency

Controlling frequency can directly influence the power of the ASIC. In terms of functionality, a designer may want to run the chip at the highest speed possible. However, higher speed generally translates to higher power dissipation. For power-sensitive designs, the designer would evaluate the speed vs. power trade-off and select the frequency of the design which meets the power constraint.

In idle mode, significant power savings can be achieved by lowering the clock frequencies of the ASIC. For example, at full speed, an ASIC may have three clocks 10, 100 and 500 MHz. In idle mode, the clocks can be switched to operate at lower frequencies such as 10, 100 and 500 kHz. This can have a substantial impact on reducing power in idle mode. Of course, if feasible, it is preferable to completely shut off the clocks in the idle mode.

See Fig. 6.2 for an example. The 200 and 300 MHz clocks are shut off using clock gaters (see Sect. 6.6 on architectural clock gating), while the frequency of the 10, 100 and 500 MHz domains are reduced in idle mode.

6.3   Dynamic Voltage Scaling

Normally, an ASIC is designed to meet the target speed for various process and operating conditions seen by the device. Thus, the designer ensures that the *slowest* part can meet the target frequency at the lowest power supply that can be seen by the device (for the entire range of junction temperature specified). Generally, the *slow*

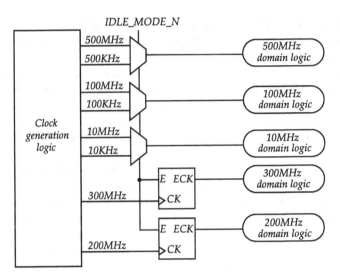

Fig. 6.2 Reducing frequency or shut-off clocks in idle mode

parts have higher performance as the temperature increases.[1] The performance of the *slowest* part would thus exceed the target frequency when the power supply value and/or the temperature is higher than the minimum possible. For the *typical* or *fast* parts (or the parts which are not from the *slowest* possible wafer), the performance would be better than the target frequency at all conditions. Similarly, for the *slow* parts when operating at minimum power supply, the performance will be better than the target frequency as the temperature is increased.

Unlike the *normal* ASIC flow described above, the dynamic voltage scaling technique adjusts the power supply voltage dynamically. In this technique, the voltage of the chip is adjusted dynamically to provide the required power and/or performance.

On-chip monitors provide feedback to either on-chip or off-chip voltage regulators to dynamically change the chip voltage. Figure 6.3 shows such a scenario where the voltage regulator is outside the chip.

A speed monitor is placed on-chip to provide feedback on the achievable speed under the specific operating conditions of the device. The speed monitor can be a simple structure such as a ring oscillator. The oscillation frequency of the ring oscillator depends upon the process condition[2] of the ASIC as well as the power supply voltage and temperature. By using the speed monitor, the power supply to the ASIC is adaptively adjusted so that it meets the target speed requirements. Figure 6.4

---

[1] This assumes temperature inversion, so that the delays for slow process condition increase as the temperature is reduced.

[2] For example, whether the device corresponds to slow process, fast process (or any condition in between).

Fig. 6.3 Off-chip voltage regulator controls chip power supply

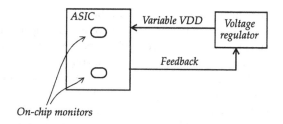

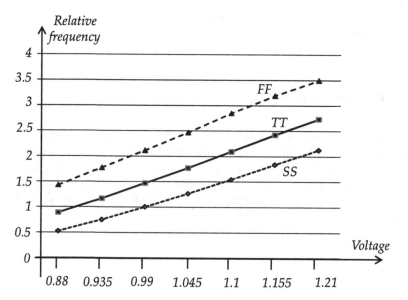

Fig. 6.4 Voltage scaling for different process conditions

shows an example of the variation of the speed with power supply for a few sample process conditions for 45 nm technology. The horizontal axis is the power supply voltage and the vertical axis shows the relative frequency that is achievable for different process conditions. The nominal power supply voltage is 1.1 V and thus the relative frequency is normalized with respect to the *slow* process, minimum voltage (10% below nominal).

This technique is also referred to as adaptive voltage scaling (AVS).

6.4   Dynamic Voltage and Frequency Scaling

Dynamic voltage and frequency scaling (DVFS) dynamically adapts voltage and frequency for different blocks. Look-up tables or on-chip monitors are used to adjust voltage and frequency depending upon performance requirements. Speed degradation or reducing the frequency of a block allows a lower voltage to be applied, which results in lower power. Alternately, a higher operating voltage leads to higher power and allows for increased performance. Notice that power is a function of the square of the voltage. Thus, reducing the voltage has a significant impact on power.

A design would typically have multiple functional modes—high speed mode and a lower performance mode. The design implementation is targeted for high performance. During the time the design is not operating at full performance, the power supply is reduced to account for reduced performance.

Significant power savings can be achieved using this approach, though it is expensive in terms of architectural design, verification and implementation. The impact on area and timing is minimal.

6.5   Reducing VDD

Since power is proportional to square of voltage, it may be beneficial to operate different blocks at different voltages. As described in Chap. 5, different portions of the design can operate in different power domains. Blocks that do not need high performance can operate at lower voltages and blocks that demand higher performance can operate at higher voltages. The voltages for the blocks may be static, that is, they need not change dynamically.

Of course, if a block can be completely shut down when not needed (*VDD* is switched *off*), this can also lead to substantial power savings.

6.6   Architectural Clock Gating

Architectural clock gating refers to the technique of adding clock gates at the architectural level to shut down the clocks of major portions of a design. Such clock gates are explicitly instantiated by the designer—as opposed to automatic inferencing by the implementation tool—and the enable signals to such clock gates are usually static[3] (see Fig. 6.5).

Architectural clock gating can also be applied at the block-level where each block in a design has a clock gate that can be used to shut down the clock to the block when not needed (see Fig. 6.6). Clock gaters can also be hierarchical, with the top-level clock gaters controlling multiple blocks, while the clock gaters within

---

[3] The enable signal being static implies that the gating control logic is dependent upon the functional mode of operation of the block and the enable signal would not change state unless the functional mode changes.

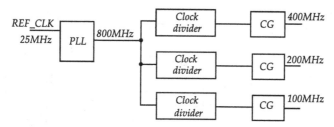

Fig. 6.5 Clock gaters on clock domains

Fig. 6.6 Multiple levels of
clock gaters

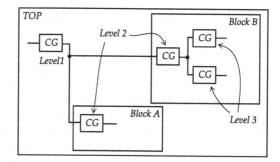

blocks control the logic just within that block. There can also be an additional level of clock gaters that turn off certain functionality within a block. Typically, architectural clock gaters do not have any impact on timing and are one of the most effective ways to save power. However, architectural clock gaters can complicate clock tree synthesis and can result in large clock skews if not carefully implemented and handled during clock tree synthesis. Of course, an intimate knowledge of the design is required to determine where such clock gaters have to be placed to get the maximum power savings.

Architectural clock gating is also sometimes referred to as coarse-grain clock gating. Fine-grain clock gating refers to the automatic inferencing of clock gates that is described in the following chapter.

6.7 Power Gating

In this technique, the supply voltage to a block (group of cells or a module) that is not active or not in use can be turned off. As described in Sect. 5.3, a header cell can be used to disconnect the *VDD*, or a footer cell can be used to disconnect *VSS*. Both ways (using the header or the footer switch) have the same effect—the power to the block is switched *off* using a control signal. As shown in Figs. 5.6 and 5.7, the *NSLEEPIN* (or *SLEEPIN*) signal controls the gating of the power (or ground).

Fig. 6.7 Single processor vs. multiple processors

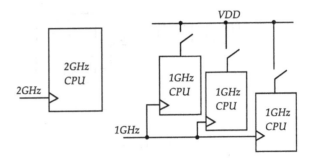

A system designer may choose between using a high throughput processor (e.g. a processor core operating at 2 GHz) and an alternate design which relies upon multiple lower speed processors (e.g. three processor cores with each operating at 1 GHz). The two approaches are shown in Fig. 6.7. For the system using three processor cores, activating all three processors can produce a throughput which may exceed that of the implementation with a single 2 GHz processor core. Using multiple processor cores has the advantage of the ability to turn off a processor during the times when the system requirements are below peak. Shutting off power to the processor cores when not required results in power (and energy) savings. In addition, since the processor cores need to operate at a lower speed (i.e. 1 GHz), the physical design for a given technology node is easier.

When the ASIC contains blocks which can be shut down along with blocks that are always *on*, there are a few considerations that the designer needs to plan in the architecture.

(a) The power to the logic that generates the control signals for shutting off various blocks should always be *on* (that is, must not be turned off). This is to ensure that the control logic for shutting *on* or *off* is always operational and does not get switched *off* inadvertently. The always-on cells have been described in detail in Sect. 5.3.

(b) Isolation cells should be placed at the outputs of switchable blocks (which can be shut off) so that a *valid* logic signal is presented to the *on*-domain (even when the switchable block is turned *off*). The isolation cells have also been described in detail in Sect. 5.3.

6.7.1   State Retention

It is possible to save the system state in a shutdown domain. The system state refers to the state of the flip-flops as well as the contents of memory instances within the block. Saving system state allows for a faster recovery when the shutdown domain is powered back *on*. Restoring the saved state avoids a complete reset at power-up, thus reducing the reset delay and the corresponding power consumption.

Fig. 6.8 Saving memory states

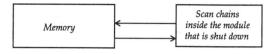

Fig. 6.9 An SRPG flop

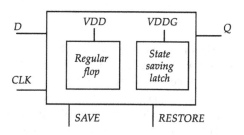

If needed, the state retention can follow one of the many scenarios described below:

(a) *System state saved externally.* Before powering down a block, the system state is saved outside of the block. The state of the system is scanned out and saved externally. When the block is powered back *on*, the externally saved state is copied back into the system. An example of saving state externally is by using an external memory (or a memory macro outside the block). To save information in a memory during shutdown, scan chains are used to connect all the flip-flops whose data need to be saved. Data can be shifted into the memory prior to shutdown (see Fig. 6.8). Upon power up, data can be shifted back into the flip-flops from the memory through the scan chains. Note that this implies that there is a time delay before the system can be powered *off* (to save its state) and another delay (to restore the state) before the system is ready for normal operation after it is powered back on.

(b) *Retention cells.* For state retention, a block can use retention cells, which can be either of a flip-flop type or a latch type. These cells are dual power supply cells and are connected to both the always-on as well as the switchable power supply. A retention cell has an additional latch, called the save-latch, that saves the state during shutdown. Figure 6.9 shows a retention flip-flop. The *VDDG* power domain is always-on and the *VDD* is the switchable power domain. In normal operation, the retention flip-flop behaves like a regular flip-flop. When *SAVE* is asserted, the flip-flop value is stored in the save-latch. Upon assertion of *RESTORE*, the value stored in the latch is restored back into the flip-flop.

Note that the *SAVE* and *RESTORE* signals should come from an always-on domain. The retention cells can only be used to save the state of flip-flops or latches; memory contents still need to be saved by one of the other methods.

(c) *Memory retention.* Similar to the dual rail retention cells, using dual rail memories allows the contents of the memory to be retained when the periphery power supply is turned *off*. Examples of dual rail memory architectures are provided in Sect. 5.3.

Table 6.1 Coarse-grain vs. fine-grain power gating

	Fine-grain	Coarse-grain
Area overhead	Large	Small
Leakage control flexibility	High	Medium
Rush currents	Small	Large

Shutting down the power of a block provides the largest amount of power saving. However, its impact needs to be considered at the architecture level (whether the power switch is on-chip or off-chip), at the verification level (to ensure that rest of the logic continues to work when the power to the block is shut down) and at the implementation level (to ensure that isolation cells are inserted and placed appropriately). The impact of shutting down a block on timing is minimal unless the critical paths go through the boundary of the block (in which case it incurs an additional delay due to the isolation cells).

6.7.2 Coarse-Grain and Fine-Grain Power Gating

The description in this book has mainly focused on power domains where a block of logic has a common power supply which can be switched off using the power switches. Such a set up is called *coarse-grain power gating* since the power switch cells control the power supply for a block of logic. The advantage of this setup is that a relatively small number of switch cells (connected in parallel) control a block of logic containing many more logic cells and also memory macros. In other words, the advantage is that the area overhead due to power switches is relatively low. However, the drawback is that the designer must keep the entire power domain *on* even when the outputs from a large portion of logic are not required.

In principle, a power switch can be placed on the power supply of every cell (or a small number of cells) enabling finer control for the shutting down of the logic. Such a set up is called *fine-grain power gating* and it can potentially offer greater power savings since the designer can turn off smaller portions of logic when those are not required to be kept *on*. The main drawback of fine-grain power gating is the large area overhead which is due to greater number of power switches and due to more complex power control and isolation logic. Managing the rush currents during the domain turning on is no longer a concern with the fine-grain power gating. The actual timing correlation with the library characterization is also improved if each cell can be characterized with the power gating transistor included within the cell.

The pros and cons of coarse-grain vs. fine-grain power gating are described in Table 6.1.

Most ASIC designs do not utilize fine-grain power gating. Unless explicitly stated, the power gating description in this book corresponds to coarse-grain power gating.

6.8   Multi-voltage

Different blocks in a design can be targeted to operate at different fixed supply voltages. A block that is timing critical can operate at a higher voltage while a block that
is not timing critical can operate at a lower voltage (see Fig. 6.10). A block that
operates at a different voltage than its surrounding logic is also referred to as a *voltage island.*

In this technique, a *level shifter* is required for any signal that is crossing from
one voltage domain to another. As described in Sect. 5.3, a level shifter cell shifts
the voltage level of the signal from the source power domain to the target power
domain. The level shifter cells are provided in the standard cell libraries in newer
technologies.

Figure 6.11 shows examples of level shifter cells placed at the inputs and outputs
of a block which is operating at a higher supply than the rest of the design. Ordinarily,
a level shifter has two power supplies,[4] one for the source power domain and one for
the target power domain. A high-to-low level shifter is used for a signal going from
a higher voltage domain to a lower voltage domain. A low-to-high level shifter is
used when the source voltage domain is lower than the target power domain.

Fig. 6.10 Different blocks
operate at different fixed
voltages

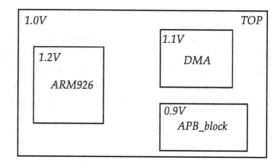

Fig. 6.11 Level shifters are
needed between two voltage
domains

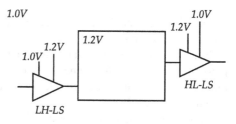

---

[4] A high-to-low level shifter may operate with only one power supply (of the target power domain).

Fig. 6.12 Level shifters for
nets going through top level

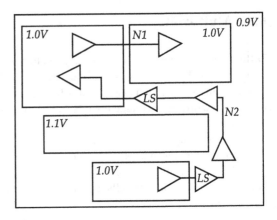

As described in Sect. 5.3.2, a level shifter ensures that the signals presented to each block have a valid voltage level. This is to ensure that timing analysis computes the delay values correctly, as well as to prevent reliability issues due to potential high leakage currents which can occur when the level shifters are not inserted.

Significant power savings can be achieved by selecting appropriate voltage values for different blocks. However this requires a lot of system design effort at the architectural level (how to select the various voltage levels, how to ensure that signal voltage levels stay within the bounds specified). There is also additional effort during verification (to ensure that logic in blocks with different voltages interact properly) and during implementation (to ensure that appropriate level shifters are inserted).

6.8.1 Optimizing Level Shifters

Level shifters are not required in every case where a net travels from one voltage domain to another. Consider Fig. 6.12.

Net *N1* has its source and target in two different power domains; however the two power domains have the same voltage. In this case, assuming that no cells are required to fix electrical design rules in the always-on 0.9 V region, no level shifter is required for net *N1*.

Level shifters are required for net *N2* even though the voltages of the source and target power domains are identical. This is because buffers are likely to be required in the always-on 0.9 V region to meet design rules as it travels from the source domain to the target domain.

Fig. 6.13 Optimize isolation cells between *PD1* and *PD2*

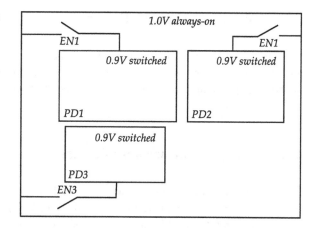

Fig. 6.14 Power domains shutting down in a sequence can reduce isolation cell insertion

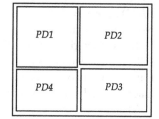

6.8.2 Optimizing Isolation Cells

It is not necessary that isolation cells get added on every signal to and from a switched domain. Consider Fig. 6.13.

The two power domains *PD1* and *PD2* are adjacent to each other in layout and the power switch control signal *EN1* is the same for the two domains. In this case, any net that traverses from *PD1* to *PD2* or vice versa does not need an isolation cell. However if the power switch controls for the two domains are different, then isolation cells between these two domains are required.

What about the case when *PD3* is shut down only if *PD1* is on and *PD3* does not come on when *PD1* is shut down. In this case, isolation cells are not required for nets going from *PD1* to *PD3*, but are required for nets going from *PD3* to *PD1*.

If *PD1* and *PD3* are situated further apart in the layout, buffers may be needed for any nets to traverse from *PD1* layout to *PD3* layout. In such cases, isolation cells are required as nets from *PD1* go through an always-on region before going back into *PD3*.

Isolation cells can be optimized based on the power state table and power sequencing information. Consider another example in Fig. 6.14.

If the power sequence for shut down is always *PD4* before *PD3* before *PD2* before *PD1*, then isolation is required only for nets going from *PD4* to *PD3*, *PD3* to *PD2* and *PD2* to *PD1*. No isolation cells are required for nets that go from *PD1* to *PD2*, from *PD2* to *PD3*, and from *PD3* to *PD4*.

6.9 Optimizing Memory Power

In the previous chapter, we have described some techniques for saving the leakage power of the memory macros during inactive mode. In this section, we describe methods for reducing the dynamic power of the memory macros.

6.9.1 Grouping the Memory Accesses

Consider two alternate scenarios where a single port memory macro is accessed (has a *READ* or a *WRITE* operation) in one-half (or 50%) of the clock cycles. In the scenario shown in Fig. 6.15, the sequence of operation is as follows. A clock cycle has a *WRITE* operation followed by an idle clock cycle, which is followed by a *READ* operation followed by another idle cycle. This pattern repeats itself. In other words, the memory macro is used with a burst length[5] of 1.

In the scenario shown in Fig. 6.16, there are two consecutive *WRITE* operations, followed by two consecutive idle cycles, then two consecutive *READ* operations, followed by two consecutive idle cycles and so on. In other words, the memory macro is used with a burst length of 2.

The clock activity contribution to the memory macro power is identical in both of the scenarios above. In each scenario, 25% of the active clock edges result in *READ* operation, 25% of the active clock edges result in *WRITE* operation and the remaining 50% active clock edges correspond to *IDLE* cycle (or no memory access). However, the *ME* and *WE* signals have 50% smaller activity in the latter scenario (Fig. 6.16) in comparison to the former scenario (Fig. 6.15). The lower activity on the *ME/WE* signals results in lower values of total dynamic power for the memory macro.

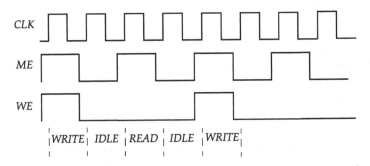

Fig. 6.15 Clock, memory enable and write enable signals for a burst length of 1

---

[5] Burst length here means the number of consecutive clock cycles with memory accesses.

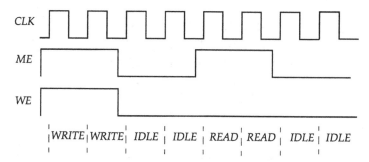

Fig. 6.16 Clock, memory enable and write enable signals for a burst length of 2

Assuming the average activity of the memory macro can be kept unchanged, increasing the burst length in memory accesses reduces the average dynamic power. Note that the average power over a smaller time interval may indeed be higher due to consecutive *READ* or *WRITE* operations.

6.9.2   Avoiding Redundant Activity on Enable Pins

Consider the example of a 2-port register file memory macro as shown in Fig. 6.17. This memory macro has one read port, *A*, and one write port, *B*. The *WRITE* operation to the memory is controlled by two control signals—memory enable *MEB* and the write enable *WEB*. Connecting both *MEB* and *WEB* pins of this memory macro to a common signal controlling the *WRITE* operation, as in scenario-*1* of the figure, results in activity on both inputs and results in unnecessary power dissipation.

Alternately, using a static signal for *MEB* and using only the *WEB* signal for controlling the *WRITE* operation, can result in savings in dynamic power. This is shown as scenario-2 in Fig. 6.17. In some cases, especially in scenarios that use small burst lengths, significant savings in the dynamic power of the memory macro can be achieved.

6.10   Operand Isolation

Data path switching, such as in adders and multipliers, can contribute substantially to switching power. However, it is possible that in some cases the data path computed values do not get used. In these cases, it should be possible to control the computation to only when necessary thus avoiding the unnecessary switching power.

Consider Fig. 6.18. If the *SEL* signal is not active, the multiplier output is not used—and the power due to unnecessary switching activity in the multiplier can be

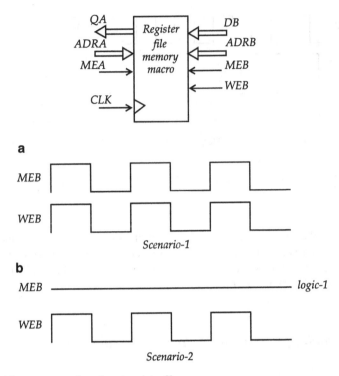

Fig. 6.17 Memory macro for a 2-port register file

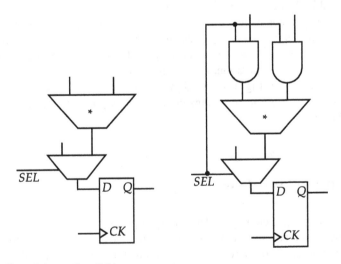

Fig. 6.18 Control data path switching

Fig. 6.19 Disabling
switching activity on
inactive path

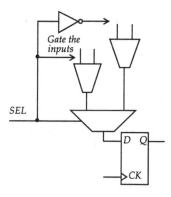

avoided. The *SEL* signal can be used to gate the input operands to the multiplier and enable these inputs only when the multiplier path is selected.

This technique can generally be applied to any complex combinational data path that is controlled by a multiplexer whose output is saved in a flip-flop. Figure 6.19 shows a case where two alternate data paths end up at a multiplexer whose output is saved in a flip-flop. The multiplexer select line can be used to disable the operands of the data path that is not actively being used.

One disadvantage of this approach is that the *SEL* signal now has to be available earlier. A better option to gating the inputs early is to hold the operand values in their flip-flops using clock gating.

6.11   Operating Modes of Design

A design can operate under multiple modes, many of which can save power. One extreme option is where the entire design is running full speed at the highest temperature. This is the worst power scenario. Power can be saved by partitioning the design based on functionality, such that only functions that are needed are powered on and the rest of the functionality is powered off.

The powering off for the functional logic implies that the power supply to the block is shut off. If that is not possible, an alternate method of saving power is to shut down the clocks, which eliminates all activity and brings the active power to zero. Alternately, non-required functional logic can run at reduced speed thus using less power.

A design may also be put into sleep or hibernate mode, for example, where all the clocks are shut down and wait for a combinational signal interrupt which wakes up the device.

In cases where certain functionality is powered down by disconnecting the power supply, it may be necessary to perform logic retention, that is, save the state of certain flip-flops or memory. This adds to the overhead during wake-up time, since the state will need to be loaded back, and thus in such cases the wake-up time is not instantaneous.

6.12   RTL Techniques

The techniques described in this section focus on minimizing the amount of logic and the number of transitions.

6.12.1   Minimizing Transitions

One should write HDL code such that data transitions are minimized, especially on a bus. For example, the HDL should not keep putting new values on a bus, unless the receiver logic is ready to receive the data. Here is an example of unnecessary data transitions on a bus and how it can be improved to reduce the number of transitions.

```
always @(posedge reset or negedge ahb_clk)
  if (reset)
    ahb_dbus <= 32b'0;
  else
    if (bus_ready)
      ahb_dbus <= read_dbus;
    else
      ahb_dbus <= 32b'0;

// Rewritten with fewer transitions:
always @(posedge reset or negedge ahb_clk)
  if (reset)
    ahb_dbus <= 32b'0;
  else
    if (bus_ready)
      ahb_dbus <= read_dbus;
// Not necessary to put default value on bus
// when bus is not ready.
```

6.12.2   Resource Sharing

Enable resource sharing on non-critical paths to ensure that minimum logic is used. Here is an example of where resource sharing can help.

```
always @(a or b or c or d or sel)
  if (sel)
    result = a * b;
  else
    result = c * d;
```

Resource sharing would create only one instance of the multiplier thus leading to power savings.

6.12.3 Others

6.12.3.1 Optimized Logic

Ensure the logic is optimized only to the extent that timing requirement needs to be met. Try to recover as much area as possible. Also eliminate redundant logic such as flip-flops with constant values. Less logic implies less power.

6.12.3.2 State Machine Encoding

One-hot and Gray code encoding lead to fewer transitions in the state register than binary encoding.

6.12.3.3 Counters

Avoid free-running counters. Try to have a start and stop requirement so that counting is bounded.

6.13 Summary

This chapter described power saving techniques that can be utilized during architectural design phase of the ASIC. Various considerations such as dynamic voltage scaling, dynamic frequency scaling, use of multiple power domains are available for low power design.

Use of architectural level clock gating along with power gating can be used to save power. Techniques such as operand isolation and HDL coding guidelines which minimize transitions are described.

Chapter 7
Low Power Implementation Techniques

This chapter describes the techniques and practices for achieving a low power implementation of the design. These implementation techniques provide power savings over and above those achieved by using the appropriate low power architecture. There are various techniques available and this chapter provides a sampling and an overview of the existing approaches.

7.1 Technology Node and Library Trade-Offs

Section 6.1 described the overall system level trade-offs impacting the selection of appropriate low power technology node. An example of a trade-off is whether the total functional power (with significant active power) should be optimized or whether the power in standby mode (mainly leakage power) should be optimized. Optimizing for total functional power may dictate a process technology node with a lower operating supply (e.g. 0.9 V) which may have a higher leakage. On the other hand, optimizing for standby power may dictate selecting a *low power* process which may have a higher operating supply (e.g. 1.2 V) but a much lower leakage. The technology nodes labeled as *low power* normally imply very low leakage. However, the low power nodes normally require a higher operating supply which increases active power.

The above is just an example of the trade-offs involved. When trying to optimize area, speed and power of a design, there are trade-offs in sacrificing one objective for the other. Optimizing for one objective only may provide poor quality of result. For example, optimizing for speed only would likely result in a design with an unacceptably large power consumption.

The overall objective—whether peak or average power, functional or standby power etc.—is a key criteria in the selection of the process technology node used for implementing a design. Once the technology node and the appropriate libraries are chosen, the techniques described in this chapter are used to obtain a low power implementation for the design.

R. Chadha and J. Bhasker, *An ASIC Low Power Primer: Analysis,*
Techniques and Specification, DOI 10.1007/978-1-4614-4271-4_7,
© Springer Science+Business Media New York 2013

7.2 Library Selection

This section examines multi-Vt and multi-channel cells.

7.2.1 Multi-Vt Cells

A multi-threshold library contains cells with different threshold voltages for MOS devices. The standard cell libraries provide more than one flavor of cells, each with a different power speed characteristic, which is typically determined by the Vt (threshold voltage) of the transistors used in the cells. For example, a library may contain High Vt (HVt), Standard Vt[1] (SVt) and Low Vt (LVt) classes of cells. The HVt cells have a higher Vt, lower leakage and dynamic power, and have larger delays. The LVt cells have a lower Vt, higher leakage and dynamic power, and have smaller delays. The corresponding characteristics of SVt cells are in the middle.

Consider the data in Table 7.1 for a representative multi-threshold library with three different Vt classes. While Vt class 1 is the lowest Vt, it provides the fastest performance at the cost of highest leakage. Vt class 3 is the highest Vt, provides the lowest leakage but offers slowest performance.

As shown in the table, while the leakage power shows a large variation due to Vt, the variation in the active power due to Vt is much smaller.

The advantage of using such classes of cells is that it is easier to trade-off power and speed where required. For example, on critical paths, LVt cells can be used, and on non-critical paths, HVt and SVt cells can be used to save power. Table 7.2 shows a typical trade-off between these three classes for a 55 nm library.

Table 7.1 Multi-Vt classes

Vt	Leakage power (nW)	Active power (nW/MHz)	Speed (MHz)
LVt (class 1)	1,000	6	400
SVt (class 2)	250	5.5	300
HVt (class 3)	40	5	250

Table 7.2 Speed power trade-off between Vt classes (relative comparison)

	HVt	SVt	LVt
Basic *nand* cell delay (slow corner)	20	16	14
Leakage power (max leakage corner)	30	60	200

---

[1] Standard Vt is sometimes referred to as Regular Vt.

The disadvantage of using multiple Vt cells is that each additional Vt type adds extra processing steps and mask cost for manufacturing. In addition, the ASIC design tools used also need to support the automatic trade-off of these cells—to enable selection between multiple Vt cells where required based on cost specifications for the design.

A design may use a mix of cells with different Vt classification. Clearly, use of high Vt cells helps reduce the leakage power. In terms of active power, the low Vt cells allow higher performance and have higher active power per cell. It is however possible that a high performance design if implemented with HVt cells may require many more cells and thus result in higher active as well as total power. This is especially true for designs which not only use higher clock frequency, but have high switching activity for signal nets. The trade-off between using standard or lower Vt cells for implementation depends upon the activity in the design, performance of the design, as well as the technology node in which the design is implemented. The intent here is to make the reader aware of the trade-offs involved in implementation to achieve a low power design.

7.2.1.1   Optimization Approach: Leakage Power Recovery

A common approach is to do an initial implementation using LVt cells. After the target performance is achieved using LVt cells, other tools and techniques can then target leakage power reduction by minimizing the use of lower threshold cells. That is, the cells on the non-critical timing paths are switched to standard (or high) threshold (that is, SVt or HVt) cells. The LVt cells then remain on the timing critical paths only. As described previously, the multi-Vt cells are typically designed to have the same footprint. Thus, in-place replacements can easily be performed post-timing closure.

A variation of the above approach involves reducing the strengths of cells on non-critical timing paths till those become critical. Note however that this approach is generally part of the timing optimization within the automated place and route tools.

7.2.2   Multi-channel Cells

Most of the standard cells in a library are built using a fixed channel length for the MOS devices. Normally, the channel length used is the minimum supported for the technology node so that the cells provide the best performance achievable within a given silicon area. This has been the norm in almost all the technology nodes. However, as the technology nodes shrink from 40 to 28 nm and down to 22 nm and finer technologies, the leakage power becomes a significant criterion for the library selection. This has caused the library vendors to provide standard cell libraries using the non-minimum channel lengths also. For example, for a technology node where 40 nm is the minimum length, standard cell libraries using 44 nm, 48 nm, etc. for MOS device lengths may be available. Since the MOS device lengths are longer

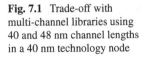

Fig. 7.1 Trade-off with multi-channel libraries using 40 and 48 nm channel lengths in a 40 nm technology node

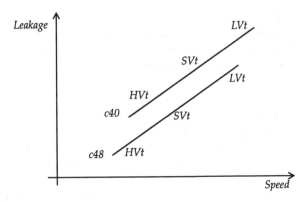

than the minimum allowed by the technology node, these are referred to as *longer channel length libraries*. The longer channel libraries normally incur a slight area penalty, that is, standard cells built using longer device lengths may be slightly larger than the minimum length standard cells.

In terms of performance, MOS device current is lower as the device length is increased. Thus, the longer channel libraries have lower performance and leakage in comparison to the standard cell libraries which are built using minimum length. For example, in a 32 nm technology node, the libraries can be available for 32, 35 and 38 nm channel lengths and a design can use cells from all of these libraries—called *multi-channel libraries*.

It should be noted that the use of multi-channel libraries allows the designer to trade-off performance with leakage similar to the scenario using multi-Vt libraries described in previous subsection. One critical difference is that using longer channel length libraries does not normally require additional wafer processing or mask cost. Note that the multi-channel libraries can be built using different Vt variants. For the above example, the 32, 35, and 38 nm channel length libraries can be available in HVt, SVt and LVt flavors, thus providing nine combinations (three Vt for each channel length). Figure 7.1 shows the trade-off between speed and leakage for multi-channel libraries in a 40 nm process technology node.

In summary, the multi-channel libraries offer similar trade-offs as available from multi-Vt libraries. The multi-Vt and multi-channel are often combined so that the design can have SVt multi-channel, in addition to HVt multi-channel or LVt multi-channel libraries allowing a finer control of the power/performance trade-off.

7.3   Clock Gating

Clock gating is a commonly used approach to reduce the dynamic power. In a typical design, a significant proportion of the dynamic (or active) power is spent due to clock pin switching at the input of each flip-flop. This section describes the clock gating approach to reduce this component of the dynamic power automatically.

Fig. 7.2 Flip-flop with an
enable

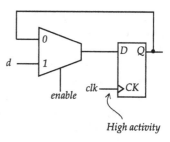

High activity

Fig. 7.3 Flip-flop with clock
gating

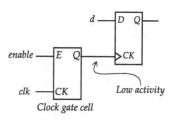

Clock gate cell

In general, a flip-flop does not capture a new value on every clock cycle. Quite
often, the output of the flip-flop is fed back to the input which is then saved on
cycles where the flip-flop's value needs to be retained. However, this causes the flip-
flop to be clocked on every clock cycle and thus consume more power than necessary. Clock gating technique provides a mechanism to shut off the clock to the
flip-flop when the flip-flop's value needs to be retained. This technique can lead to
substantial power savings depending upon how often a new value is latched into the
flip-flop. On an extreme, if a flip-flop is being clocked with a new value on every
clock cycle, then gating the clock is ineffective. However if a flip-flop latches with
a new value once in 100 cycles, then 99% of the clock cycles can be gated and 99%
of the flip-flop clock power can be saved.

Consider the following SystemVerilog[2] code.

```
always_ff @(posedge clk)
  if (enable)
    q <= d;
```

This would typically be implemented as logic shown in Fig. 7.2.

Notice that when *enable* is false, the Q output is fed back to the input D which is
flopped again. Such feedback loops cause unnecessary increase in power. This extra
power can be saved by not providing the clock signal when *enable* is false so that
the flip-flop preserves its previous value. This can be accomplished by using a clock
gater as shown in Fig. 7.3. The clock gate prevents any clock edges from getting to
the flip-flop while *enable* is false.

---

[2] See [BHA10].

The clock gating illustration on the flip-flop in Fig. 7.2 is an example of a synchronous load-enable. The same transformation can be applied to flip-flops that have a synchronous set or a synchronous reset.

Clock gating is helpful in reducing power. It further eliminates the need for a multiplexer on the data path and thus makes it easier for these data paths to meet timing and results in saving silicon area. Lower toggle rates on the clock pin implies lower internal power for the flip-flop. Since the clock gate itself contributes to power, typically a clock gate is used to drive a minimum number of flip-flops. This ensures that the power saved by gating the flip-flops is larger than the power dissipated due to the clock gater.

A possible timing issue is that the data path timing to the enable pin of the clock gater may become critical due to the insertion of clock gates. The timing impact of clock gates is described in Sect. 7.4.

The present day synthesis tools automatically infer the synchronous controls from an RTL description and map them to use clock gaters from a technology library. This technique is easy to implement, is technology independent and requires no RTL changes. Some tools may even take a netlist and find structures (multiplexer at data input of flip-flop with output of flip-flop going back to multiplexer) that can be replaced with clock gaters.

One option specified when inferring the clock gaters is the minimum number of flip-flops that a clock gater can drive. Driving only one flip-flop may not be economical in terms of power, as the power savings may be nullified by the power dissipated by the clock gater. Typically, a minimum fanout of 3 or 4 is used for each clock gater.

During physical design, clock gate cells are placed closed to the flip-flops being gated—this reduces capacitance and eases the timing requirement for the enable pin of the clock gater. Note that the part of the clock tree beyond the clock gater becomes part of the timing path going through the enable pin. Clock gates can be cloned and pushed higher in the clock tree (that is, closer to the flip-flops) to achieve better clock skew—this also eases the constraints on the enable pins. Alternately, at the cost of making timing closure on enable pins more difficult, clock gates can be merged and pushed lower in the clock tree, gating larger number of flip-flops and leading to better power and smaller area. See Sect. 7.4 for more details.

7.3.1 Power-Driven Clock Gating

Knowing the activity of a clock gater can help determine the trade-off between power savings and the cost of putting in the clock gaters. The activity of a clock gater can be obtained from a simulation and provided in an activity file, such as SAIF. If a clock gate is active most of the time, it may be better not to have a clock gater and to use the conventional multiplexer scheme. Optimizing the selection of clock gaters based on activity is known as *power-driven clock gating*.

7.3.2   Other Techniques to Reduce Clock Tree Power

This section discusses useful skew and flip-flop clustering techniques.

7.3.2.1   Useful Skew

In general, a clock tree implementation attempts to achieve the smallest clock skew at all the leaf flip-flops (termination nodes of the clock tree). To understand the concept of useful skew, we first describe the typical steps utilized during the clock tree implementation.

The first step involves building a tree to all the leaf nodes. Since the insertion delays to various leaf nodes is, in general, quite different (say ranging from *Tmin* to *Tmax*), the next step is to balance the insertion delays to various leaf nodes. This step adds repeater stages in various branches so that the delay to all the branches is ~*Tmax*. Note that the additional repeater stages (for skew balance) can be a significant fraction of the clock tree. In other words, a significant portion of clock tree power is consumed by the repeater stages (can be inverters or buffers) added only for skew balancing.

Useful skew implies that the clock tree is not balanced at the leaf nodes. Here the goal is not to make sure that all endpoints of a clock tree have identical latency, but to ensure that only the necessary skew buffer stages are added to meet timing. An intentional skew may be added between specific leaf nodes that require longer timing paths (over and beyond one clock cycle) simplifying also the timing closure. Instead of balancing the clock at all leaf nodes, the clock at some leaf nodes is *pushed out* or alternately *brought forward*. Thus, unlike a normal clock tree implementation[3] where all data path stages have nearly equal time available, a useful skew implementation provides different stages of data path with significantly different available times. The available time for a data path may be more (or less) than one clock cycle depending upon the amount of useful skew. Thus, a significant power reduction can be achieved for the implementation that uses useful skew.

Consider Fig. 7.4. Even though the clock period is 5 ns, adding a useful skew (difference between the clock arrival times at the flip-flop clock pins) of 2 ns allows this path to meet the setup timing requirement. If the clock skew were to be 0, additional buffers would have to be inserted in the clock branch to *FF0* thus increasing power, and further data path optimization would have to be accomplished to meet the setup timing (since only 5 ns would be available for the data path).

The disadvantage of utilizing useful skew is that you may end up with a larger number of hold buffers, plus have a larger timing impact from OCV[4] margin due to larger non-common clock paths.

---

[3] Built with minimum clock skew.
[4] On-Chip Variation.

Fig. 7.4 Useful skew to meet
timing

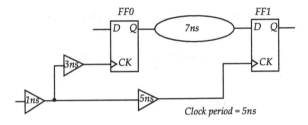

7.3.2.2   Register Clustering

By keeping all the flip-flops that fan out from the last clock tree buffer close together, the clock buffer can be placed in the same cluster which thereby reduces capacitance and thus reduces dynamic power (which is proportional to capacitance).

7.4   Timing Impact Due to Clock Gating

During implementation, the use of clock gating imposes additional timing constraints. This section describes the impact of clock gaters on timing.

7.4.1   Single Stage Clock Gating

One of the factors to be considered for clock gating is that it can introduce tight timing paths, mainly to the enable pins of the clock gaters. Consider Fig. 7.5.

Assume that the clock period is P. Without a clock gater, the setup requirement at *FF1* is:

```
P > launch_clock_insertion_delay + CK_to_Q_of_FF0 + T +
    SetupTimeFF1 - capture_clock_insertion_delay
```

With a clock gater in the clock path of the capture flip-flop *FF1*, the capture clock insertion delay increases by the *CK_to_Q* delay of the clock gater. Further, a multiplexer at the *D* input of *FF1* no longer exists. Both of these factors actually help in the setup timing for *FF1*. However, the enable logic has moved to the enable pin of the clock gater. If *T2* is the latency from the clock gater to the flop *FF1*, then the available time for the enable logic to be captured by the clock gate is smaller by $(T2+CK\_to\_Q\_of\_CLKGATE)$. This implies that the timing path *T1* must be faster than *T* by the amount $(T2 + CK\_to\_Q\_of\_CLKGATE)$. What this means is that *T1* (which is the path delay to *E* pin) needs to be made smaller than a full cycle so that the timing to *E* pin can be met.

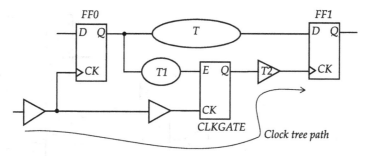

Fig. 7.5 Tight timing paths to enable pin

Fig. 7.6 Large fanout of a clock gater

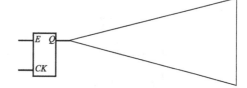

During synthesis, clocks are considered as ideal. Therefore, setup timing to E pins of clock gaters need to be met with an additional margin (based upon an estimation of T2). This can be specified using *set_clock_latency* commands of, say, −400 ps (an estimated value of T2) on the *CK* pin and +400 ps (the estimated value of T2) on the Q pin of the clock gater. If the clock insertion delay to *CK_of_FF1* is *INS_DELAY1*, the latency to the *CK* pin of the clock gater can be set to (*INS_DELAY1 - T2 - CK_to_Q_of_CLKGATE*). During physical design, care must be taken to ensure that the clock gaters are placed close to the flip-flops that are being gated. Otherwise, timing failures to the enable pins of the clock gaters are likely.

The closer a clock gating cell is placed to the flip-flops that it drives, the less constrained the corresponding enable signal becomes. Let us now look at the impact of fanout on clock gaters. Figure 7.6 shows a scenario of a clock gater driving a large number of flip-flops. This approach uses fewer clock gating cells and has better power reduction. However, the enable pins of the clock gaters are heavily constrained (since there is a large latency from the output of the clock gater to the flip-flops being driven).

Consider the alternate scenario shown in Fig. 7.7 where each clock gater drives fewer flip-flops. In this case, it is easier to meet the enable pin timing. However, power may be impacted, as the number of clock gating cells is larger.

Fig. 7.7 Small fanout of a
clock gater

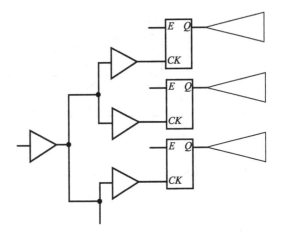

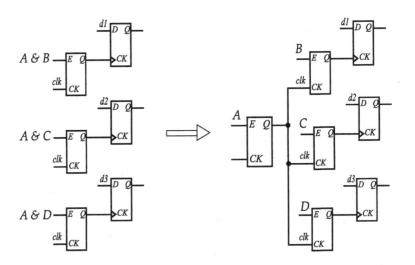

Fig. 7.8 Multiple levels of clock gaters

7.4.2 Multiple Stage Clock Gating

There can be multiple levels or stages of clock gaters. These stages can automatically
be inferred by a synthesis tool depending on the common logic in the enable pin
logic cone. Figure 7.8 shows one such transformation that can create multiple levels
of clock gaters. The common factor A in the expression of the enables has been fac-
tored out into an additional level of clock gater which has A on its enable pin.

Consider the general case when there are multiple stages of clock gaters. The
timing to the enable pins of the clock gaters—especially at the first stage—becomes
even more critical. Figure 7.9 shows an example with three stages of clock gaters.

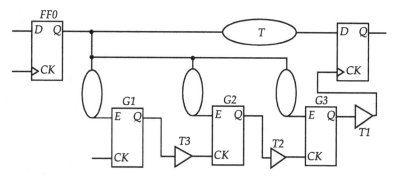

Fig. 7.9 Timing paths with multiple levels

Fig. 7.10 Hard to meet timing to a clock gate due to large *N1* latency

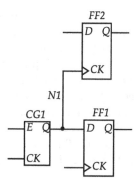

The problem with multiple stages is that now the timing to the enable pins of the clock gaters becomes more critical. The enable pin of *G1* is tighter by $(T1 + T2 + T3)$, enable for *G2* by $(T1 + T2)$ and enable for *G3* is tighter by *T1*.

7.4.3 Cloning Clock Gates

Clock gates can be cloned sometimes to help meet timing. The cloning can be for the clock gater enable pin or for the flip-flops being driven. This results in a likely increase in clock power but helps in meeting the timing requirement.

See Fig. 7.10. Consider the case where flip-flop *FF2* is physically far away from flip-flop *FF1* and clock gate *CG1* is close to *FF1* in a layout. The latency on net *N1* will be large due to the *CG1* to *FF2* distance. Assume that the latency of net *N1* is *T1*. In such a case, the timing on the enable pin of the clock gate has to be tighter by at least *T1*. The setup constraint to the clock gater has now to be tighter by *T1* and it

Fig. 7.11 Cloning clock
gates

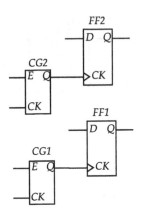

may become impossible to close timing for this path. Note that this constraint arises due to the latency of the clock from *CG1* to *FF2*, which is farther away.

In cases like this, the timing to the clock gater can be resolved by cloning the clock gater, that is providing a separate clock gater for *FF2* which is placed close to *FF2* in layout (see Fig. 7.11). This removes the tight setup constraint to *CG1*, which helps to meet the required timing. Note that this comes at a cost of using an additional clock gater, which causes more power to be consumed.

Note that another option to improve timing without doing cloning is to attempt to move the flip-flop *FF2* closer to *FF1*, for example, *FF1* and *FF2* can be clustered to be physically close to each other.

7.4.4 *Merging*

Clock gates can be merged if they provide any area or power savings without sacrificing timing. Of course, the enable pin logic must be identical to allow such merging to occur. Consider Fig. 7.11, and assume that flip-flops *FF1* and *FF2* are physically close to each other. If the logic connected to the enable pin of *CG1* is identical to the logic connected to the enable pin of *CG2*, then one of the clock gaters can be eliminated—the remaining clock gater provides the clock to both the flip-flops.

Note that it is not a necessary requirement for the flip-flops to be physically close to each other. For example, consider the case where *FF2* is physically far away from *FF1* and *CG2* has been chosen to be eliminated. It is possible that the slack on the logic driven by *FF2* is positive enough such that the *CG2* clock gate can be eliminated. After the *CG2* is eliminated, the clock at *FF2* arrives later which reduces the positive margin for the logic driven by *FF2*. This implies that this technique can be employed only when the original slack on the logic driven by *FF2* is larger than the incremental delay on the *FF2* clock path incurred by eliminating *CG2*.

7.5   Gate-Level Power Optimization Techniques

The gate level power optimization techniques require that the switching activity information on pins of gates is available.

1. These can be derived as accurate values through SAIF/simulation or,
2. These can be obtained by using an approximate activity based on SDC, or
3. These can be obtained by other means, such as explicit activity specifications.

7.5.1   Using Complex Cells

If a net has a much higher activity than its fan-in or fan-out nets, collapsing the high activity net into one complex cell can reduce the overall dynamic power. This is because collapsing reduces the capacitance (and the corresponding switching power) for the high activity net. Figure 7.12 shows an example.

7.5.2   Cell Sizing

Downsizing the cells on the non-critical paths can help reduce the dynamic power consumption. A lower strength cell has a lower dynamic power consumption. This is normally part of the area recovery and the power recovery steps during the timing and power driven implementation. In this approach, the place and route tool would reduce the cell strength of the non-critical timing paths to save area and reduce power after the initial implementation is completed.

Lower strength cells normally provide a lower input capacitance load on the previous stages than the higher strength cells. Higher strength cells are useful for driving long traces or for driving heavier loads. Lower strength cells are typically optimal for low power.

During the physical design of an ASIC, a program can target power improvement by reducing the sizes of the cells on the non-critical paths. The lower strength cells most likely are not footprint compatible. However, the lower strength cells have usually lower area than the higher strength cells and can thus fit within the existing area.

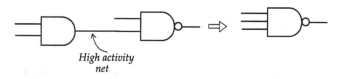

High activity net

Fig. 7.12 Technology mapping to optimize high activity net

Fig. 7.13 Swapping pin to a
lower capacitance pin

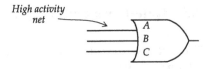

7.5.3   Appropriate Slew Target for Design

In general, having a very slow transition time (or slew) can lead to higher dynamic
power. However, targeting very fast slew can lead to too many repeater stages which
can cause congestion and can contribute to higher leakage power. Thus, the slew rate
values targeted during an implementation depends upon the frequency of operation.
In general, a good design practice is to target the clock signal transition times to be
~10% of the clock period. The transition times for the data signals can be larger than
that of the clock signals since the data toggle rate can be at most 50% of the clock
toggle rate. In most designs, the data signal transition times can be set to ~30% of
the clock period. Repeater stages are inserted as required on signal nets which have
slow transition times and violate the target slew rate chosen for implementation.

7.5.4   Pin Swapping

In certain cells, pins with equivalent functionality can have different input capacitances. In such cases, it is beneficial to move the high activity net to the pin with the
lower capacitance. See Fig. 7.13 for an example. The logic on pin *A* can be swapped
with the logic on pin *C* which is assumed to have smaller pin capacitance. The functionality remains the same and the high activity net now drives a smaller capacitive
load, thereby reducing power.

7.5.5   Factoring

Boolean factoring can be applied on a net with high activity to factor out the high
activity net. This minimizes the number of logic fanouts for the high activity net.
For example, if net *B* is a high activity net and its logic function is:

    A & B | B & C

then factoring can be applied to change it into:

    B & (A | C)

This is illustrated in Fig. 7.14. The number of high activity nets has been reduced
by logic factoring.

Fig. 7.14 Factoring high activity nets

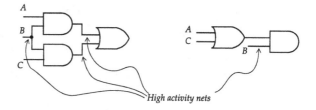

High activity nets

7.6 Power Optimization for Sleep Modes

This section describes techniques for power reduction when the design (or a portion of a design called a *block*) is not operating in the functional mode—i.e. the design (or the *block*) has been placed in the *sleep* or *standby* mode. In general, various macros such as SerDes, PLLs, DDR2/3 PHY, IOs, etc. support a power down mode which brings down the power dissipation in these macros by a large proportion.

This section describes power reduction techniques that can be implemented for the memories and the standard cell logic. Since the design is in inactive mode, there is no activity (or zero activity for various nodes) resulting in zero dynamic (or active) power for this mode.

7.6.1 Back Bias for Leakage Reduction

This technique is often employed to reduce leakage when the device is inactive.

Back bias is the technique of dynamically changing the threshold voltage (Vt) of a CMOS transistor. Increasing the threshold voltage of a transistor reduces its leakage current, but performance gets degraded. Since the design is not operating in the functional mode, the degradation in performance is acceptable since the goal is to reduce the power. Back bias is also sometimes referred to as *substrate bias*, *well bias*, *body bias* or *back gating*.

Consider the NMOS and PMOS transistors shown in Fig. 7.15. A MOS transistor has these nodes: *source, gate, drain,* and the *substrate.*

In general, the substrates of all NMOS transistors of digital standard cells are connected to *VSS*, while the substrates of all PMOS transistors of digital standard cells are connected to *VDD*.[5]

The p–n junctions within MOS transistors form diodes and substrate bias should not be applied to turn these on (that is, the diode should not get forward-biased). In the NMOS transistor example, applying forward bias (positive voltage difference) on the substrate compared to source can turn on the diode. However, any reverse

---

[5] The substrate connections can be built in the standard cells or can be through separate "tub-tie" cells.

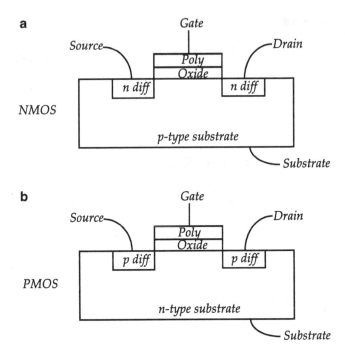

Fig. 7.15 NMOS and PMOS transistors with substrate pin

bias (negative voltage difference) can be applied on the substrate, which causes leakage current to reduce without any detrimental effect.

The threshold of an NMOS transistor is impacted by how much negative bias is applied to the substrate (which causes more electrons in the substrate that must be overcome by a higher voltage on the gate).[6] The greater the magnitude of the negative bias, the higher is the positive gate voltage required to turn on the transistor. This in turn implies lower leakage and also lower speed. Similarly, for the PMOS transistor, a positive bias applied to the substrate implies that a larger negative gate voltage (relative to *VDD*) is required to turn it *on*.

Back bias can also be used to improve yields. If a chip is consuming too much power due to leakage, then back bias can be applied to reduce the power. However, this also results in speed degradation. Alternately, a slightly positive back bias can be used to improve performance at the expense of increased leakage power.

A significant drawback of applying the substrate bias is that this technique requires a separate set of power supplies—a negative supply *VBB* for the NMOS transistors and another positive supply *VPP* (higher than *VDD*) for the PMOS transistors. Another drawback is that some tools do not support automatic techniques for back-bias connectivity.

---

[6] See [SZE81, STR05] for further details.

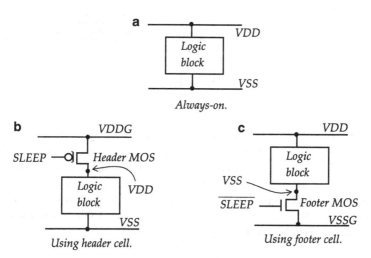

Fig. 7.16 Cutting off power to an inactive logic block using a header or a footer device

7.6.2   Switching OFF Inactive Blocks

In this method, inactive blocks are shut down to save power. By turning the blocks off, the active power goes down to zero. The control for turnoff is internal to the chip and the shutdown is accomplished using power switches.

Power gating involves gating off the power supply so that the power to the inactive blocks can be turned off. This procedure is illustrated in Fig. 7.16, where a *footer* (or a *header*) MOS device is added in series with the power supply. The control signal *SLEEP* is configured so that the footer (or header) MOS device is *on* during normal operation of the block. Since the power gating MOS device (footer or header) is *on* during normal operation, the block is powered and it operates in normal functional mode. During inactive (or sleep) mode of the block, the gating MOS device (footer or header) is turned off which eliminates any active power dissipation in the logic block. The footer is a large NMOS device between the actual ground and the ground net of the block which is controlled through power gating. The header is a large PMOS device between the actual power supply and the power supply net of the block which is controlled through power gating. During sleep mode, the only power dissipated in the block is from the leakage through the footer (or header) device.

The footers or headers are normally implemented using multiple power gating cells which correspond to multiple MOS devices in parallel. The footer and header devices introduce a series *on*-resistance to the power supply. If the value of the *on*-resistance is not small, the IR drop[7] through the gating MOS devices can affect the timing of the

---

[7] Voltage drop in power mesh rails.

Fig. 7.17 Daisy chain configuration

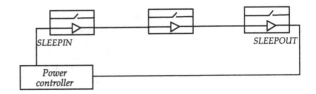

cells in the logic block. While the primary criteria regarding the size of the gating devices is to ensure that the *on*-resistance value is small, there is a trade-off as the power gating MOS devices determine the leakage in the inactive or sleep mode.

In summary, there should be an adequate number of power gating cells in parallel to ensure minimal IR drop from the series *on*-resistance in active mode. However, the leakage from the gating cells in the inactive or sleep mode is also a criteria in choosing the number of power gating cells in parallel.

7.6.2.1 Daisy Chain Configuration to Limit Rush Current

In a typical design, many switch cells are required to handle the current requirements of the switchable power domain. These switch cells can be configured in many different forms. A common form is the daisy chain configuration shown in Fig. 7.17.

The *SLEEPIN*s and *SLEEPOUT*s of the switch cells are daisy-chained. The power controller issues a signal to the first *SLEEPIN* and it receives an acknowledge signal from the last *SLEEPOUT*. The advantage of this technique is that the turning *on* of the multiple switch cells is staggered in time so that the there is no sudden increase in the *current* drawn from the power supply. A sudden increase in the *current* drawn during wake-up of this domain can place stringent requirements on the power supply. If the power delivery mechanism cannot handle the sudden spike in *current* requirements, the power supply voltage received by power domains which are already *on* can fall below the minimum acceptable level. The sudden increase in current requirements from the domain being turned *on* is called *rush current*[8] and daisy chaining of the switches helps reduce the rush current. By daisy-chaining the switch cells as shown in Fig. 7.17, the turning *on* of different switch cells is staggered so that it limits the amount of *current* spike from the power supply. The disadvantage is that the wake-up time for the shutdown domain depends on the total delay from the first *SLEEPIN* to the last *SLEEPOUT*.

7.6.2.2 Parallel Configuration to Minimize Turn-on Time

An alternate configuration for the switch cells is a parallel configuration shown in Fig. 7.18, where all the *SLEEPIN*s get their signal at the same time. In this case,

---

[8] Also sometimes referred to as *in-rush current*.

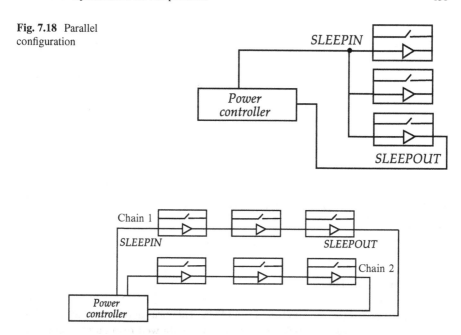

Fig. 7.18 Parallel configuration

Fig. 7.19 A parallel-serial configuration

while the wake-up time is the smallest, the current spike during wake-up may be prohibitively high. Because of the large rush current, this configuration is almost never employed.

7.6.2.3   Multiple Switch Chains to Manage Rush Currents

In this method, the switch cells are grouped into multiple chains and the daisy chain configuration is used for each chain (see Fig. 7.19). In this configuration, the control signals for different chains can be staggered (in time) to control the rush current. The first chain can control part of the switches which starts the 'ramp-up' of the power supply. The control signal for the second chain is delayed from the control signal for the first chain. Note that the mother-daughter switch cells can also be grouped into separate chains with separate controls for the mother switch chain and the daughter switch chain (see Fig. 7.20). Delaying the control signal for the mother (or subsequent) chains ensures that the rush current during power supply ramp-up is within the acceptable limits. The advantage of this configuration is that it can appropriately address the requirements of rush current and provides the control to the power controller. The *SLEEPOUT*s from the last switch of all the serial chains can be connected back to the power controller or only the *SLEEPOUT* of the later chain can be connected to the power controller. This depends upon the logic implemented in the power controller.

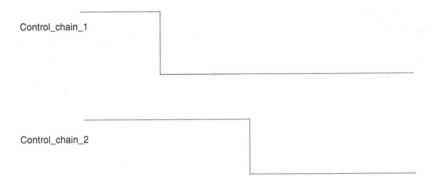

Control_chain_1

Control_chain_2

Fig. 7.20 Staggering the control signals for different chains

7.6.2.4 Wake-up Time

A critical criteria is the turn-on time for a block. The turn-on time is determined by various parameters of the block such as maximum allowed rush current, charging capacitance and how the switches are arranged in a daisy chain. Consider an example block where 5,000 power switches are used. As described previously, the number of switches is chosen to ensure that the voltage drop due to the switches is small (below the acceptable limit). Note that each switch has some propagation delay from *SLEEPIN* to *SLEEPOUT*. As an example, if each switch cell has a 30 ps propagation delay (to *SLEEPOUT*), having 5,000 switches in the chain would result in roughly 150 ns (=5,000 × 30 ps) propagation delay when all the switches are placed in one chain. If the power supply capacitance cannot be charged within 150 ns and the total current through all the switches exceeds the maximum allowed rush current (is too large), the designer may place the 5,000 switches in multiple groups which are turned *on* independently.

7.6.3 Sleep and Shutdown Modes for Memories

Power-friendly memories typically offer various power saving mechanisms labeled as sleep modes. These are broadly aimed at saving the leakage power for the peripheral logic and the core memory array. Some of the methods for saving leakage power have been described in Chap. 3. This section elaborates on the power reduction alternatives for the memory macros.

Fig. 7.21 Sleep mode turns off peripheral logic in a memory

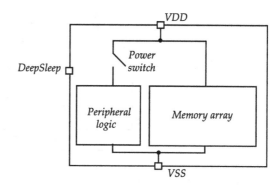

7.6.3.1 Power Savings for Peripheral Logic

Since the peripheral logic power supply does not affect the memory contents, the peripheral logic power supply can be shut down (or reduced[9]) without affecting the memory contents. The shutting down of the peripheral logic power supply can be implemented as following.

(a) *One of the sleep modes of the memory macro turns off the supply to the peripheral logic.* This is achieved normally using an internal power switch which shuts *off* power to the peripheral logic. A schematic illustration is shown in Fig. 7.21 where the power switch is assumed to be controlled by the *DeepSleep* control pin.

There is some delay before the internal power switches can be activated to remove power for the peripheral logic. Similarly, there is a significant delay after the memory *sleep* mode is deactivated until appropriate power gets applied to the peripheral logic enabling the memory macro to be active. The designer needs to consider the wake-up time while deciding to de-activate the power supply of the peripheral logic.

Depending upon the size of the memory instance, up to 50–70% reduction can be expected in static (or leakage) power in this mode. The data in the memory is still safe. However, the wake-up time is longer—to allow the power of the peripheral logic to stabilize, and this may take multiple clock cycles.

(b) *The memory macro supports separate power supplies for the peripheral logic and the memory core array.* For instances with dual power supplies, the control of peripheral logic supply can also be external. If the memory macro is part of a larger block that is a shutdown domain, the block level power supply connected to the standard cells and the memory peripheral logic can be shut down without affecting the memory core array (see Fig. 7.22). The memory core array stays powered to ensure that the memory contents are not lost.

In a memory with dual rails, the peripheral logic and the memory array can operate at different voltages. This allows the memory array power to be lowered

---

[9] Unlike the memory array power supply which must be kept above a specified minimum to ensure no loss of contents, the peripheral logic power supply can be reduced to a much lower value.

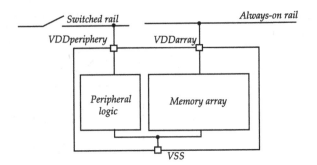

Fig. 7.22 Dual power rails in a memory

to its minimum *allowable* value and the peripheral logic to be even lower—the memory macro can still operate in functional mode and provide savings in dynamic power.

7.6.3.2   Power Savings for Memory Array

Since the power shutoff to the memory array causes the loss of memory contents—the memory core array power supply is generally not removed.[10] One way to save leakage power for the memory core array is to provide a substrate bias for the memory array. The bias generation circuitry is internal to the memory macro and the bias applied results in significant leakage reduction for the memory array. Figure 7.23 depicts a separate pin *LightSleep* to control this mode. Depending upon the memory size, adding a substrate bias can result in up to 20–50% reduction in standby power. The data in the memory is still safe and the wake-up time, that is the time to come back to normal operation, is fast and may typically be less than one clock cycle.

7.6.3.3   Memory Shutdown

The techniques described so far retain the contents of the memory. If the designer no longer needs the memory contents (since these have been saved separately), the power to the memory array (and the peripheral logic) can be shut down and the leakage power for the memory can be eliminated completely. The contents of the memory are lost in the *shutdown* mode. Thus, before placing the memory in the *shutdown* mode, the system designer must ensure that the contents are either not required for subsequent operations, or they are saved in another block (or stored off-chip). After the block is powered back *on*, the memory contents are brought back and re-written to the original locations before the block is ready for subsequent operations.

---

[10] The memory core array power supply is removed in shutdown mode where the memory contents do not need to be saved.

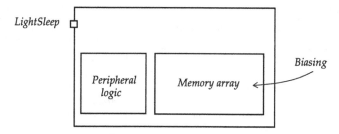

Fig. 7.23 Light sleep mode in a memory

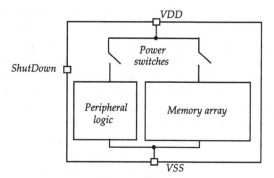

Fig. 7.24 Shutdown mode in a memory

In the shutdown mode, the power to the memory array is removed (see Fig. 7.24). A separate pin *ShutDown* controls this mode. Since the power to the memory array is also removed, the entire data stored in the memory is lost. This may normally require a large number of cycles to restore the data. Alternately, if data is not being restored, just to bring up the power to the memory array and the peripheral logic can involve a larger wake-up time, typically longer than the wake-up time for deep sleep mode. However, the power reduction can be substantial and the savings can exceed 95% of standby power.

7.7   Adaptive Process Monitor

In most cases, a device is intended to meet the timing and power specification over the range of process, voltage and temperature of operation. This implies that the implementation should meet the timing and power targets over the entire range of PVT corners—various timing sign-off corners are used to verify timing specifications and the maximum power corner (*fast* process, high supply, high temperature) for the power specification.

Table 7.3 Device performance based on process

Wafer corner	Performance at min voltage, min temperature	Power at max voltage, max temperature
Slow	Barely passing spec.	Much lower leakage power than spec.; active power also lower
Typical	Faster than spec.	Lower leakage power than spec.; active power also lower
Fast	Much better performance than the spec.	Barely meet the spec.

The timing specification is normally limited by the performance at the slow corner (*slow* process, minimum power supply values seen on the die, and perhaps lowest junction temperature allowed—to account for temperature inversion). Similarly, the power is specified by the power dissipation at the maximum power corner (*fast* process, maximum power supply, highest junction temperature). In reality, a device from the *slow* process corner would have much lower power than the specification whereas a *fast* corner device would provide much better timing than the specification. Most devices, which are close to *typical* process, would perform better than the specification for both timing (greater than the sign-off speed) and power (lower than the sign-off power). The device performance based upon their process can be described as in Table 7.3.

For devices which target very high performance and have critical constraint for power, it is normally not practical to meet the sign-off criteria for timing and power at their respective worst-case corners. Instead of using a fixed supply with a given tolerance for all devices, a solution is to adjust the power supply based upon whether it is a fast, typical, or a slow device. This is similar to the concept of voltage scaling described in Sect. 6.3. For example, a fast device can be operated at reduced power supply and still meet the target performance. The slow device can operate at the high end of the allowable power supply range. Note that the slow devices have much lower leakage (and thus have lower total power). Similarly, by operating the fast devices at reduced power supply, both the leakage and the dynamic power contributions are reduced.

The technique described above can be implemented statically whereby during the chip testing, the device is *marked* to indicate *slow*, *typical*, or *fast* and an appropriate *fuse* is blown within the device to indicate its speed category. In the field, the power supply is set based upon the *marking* of the device.

In a more sophisticated scenario, the power supply can be adjusted dynamically based upon monitoring of the chip process and operating temperature. As described in Sect. 6.3, a simple monitor can just be an on-die ring oscillator whose speed is a measure of the process, voltage and temperature on the die. The ring oscillator speed is sensed externally (off-chip) and controls the voltage regulator which sets the power supply on the die. If the measured speed is low (due to slow process and low temperature), it implies that the device performance is slow. In this case, the external power supply value is increased. An increase in power supply improves the performance of the device. Similarly if the measured speed is high, the power

Table 7.4 Comparison summary between various approaches for power reduction

Power reduction techniques	Power benefit	Timing penalty	Area penalty	Architecture impact	Design impact	Verification impact	Implementation impact
Multi-Vt opt	Medium	Little	Little	None	None	None	Low
Multi-channel opt	Medium	Little	Little	None	None	None	Low
Clock gating	Medium	Little	Little	Low	Low	None	Low
Multi-supply voltage	Large	Some	Little	High	Medium	Medium	Medium
Power shutdown	Huge	Some	Some	High	High	High	High
Dynamic and adaptive voltage frequency scaling	Large	Some	Some	High	High	High	High
Substrate biasing	Large	Some	Some	Medium	Low	Low	High

supply value can be reduced. In either scenario above, the power supply is set so that it just meets the performance specified. The power supply setting ensures that the power dissipated is much lower than the traditional approach with fixed power supply setting for all devices.

7.8 Decoupling Capacitances and Leakage

All designs implemented in nanometer technologies include significant built-in decoupling capacitances which help reduce the power supply transients on the die. In general, a designer adds as much decoupling as feasible and uses all of the open space on the die to add decoupling capacitances. Note however that while there is no active power due to decoupling capacitances, these do contribute to the leakage power. For designs where power is a rigid constraint, the designer may want to evaluate whether using *all* available space for decoupling would result in too much leakage.

7.9 Summary

This chapter described various approaches and techniques that can be adopted for achieving a low power implementation. Different approaches have differing amount of impact on the potential power savings which need to be weighed against the differing impact on timing, silicon area, design verification effort, and impact on system architecture. Table 7.4 provides a comparison summary of the various techniques.

Chapter 8
UPF Power Specification

This chapter describes the Unified Power Format (UPF) commands. The UPF specification is in ASCII format that can be used to specify low power directives that encompass all aspects during the design flow (see Fig. 8.1). The power specification provided in a UPF file can be used by simulation, synthesis, equivalence checking, physical design and physical verification.

The UPF specification is an IEEE 1801-2009 standard.[1] Prior to that, a UPF version 2.0, which was standardized by Accellera,[2] was available. The various commands that are part of the IEEE standard are described next.

8.1 Setting Scope

The **set_scope** command specifies the instance to which the active UPF scope is set. If no instance name is specified, the scope is that of the top level.

```
set_scope u_i1/u_i2

set_scope
# Set scope to top-level.
```

8.2 Creating Power Domains

The **create_power_domain** command creates a power domain with the specified name. This command can also be used to specify the list of all instances in the power domain.

---

[1] See [UPF09] in References section.

[2] www.accellera.org.

R. Chadha and J. Bhasker, *An ASIC Low Power Primer: Analysis, Techniques and Specification*, DOI 10.1007/978-1-4614-4271-4_8, © Springer Science+Business Media New York 2013

Fig. 8.1 UPF is used by all
parts of the flow

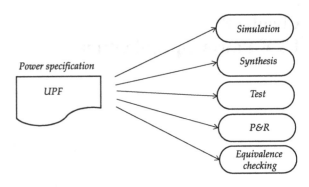

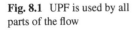

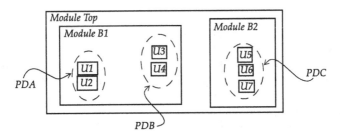

Fig. 8.2 Power domains

```
set_scope
create_power_domain PDA \
  -elements {U1 U2} -scope U_B1

create_power_domain PDB \
  -elements {U3 U4} -scope U_B1
set_scope U_B2
create_power_domain PDC -elements {U5 U6 U7}
```

The above commands correspond to the power domains depicted in Fig. 8.2. The
-include_scope option can be used to add all the elements in the specified scope to
share the same supply as the power domain (this does not correspond to the illustration in Fig. 8.2).

```
create_power_domain PDD -include_scope -scope U_B3
```

Figure 8.3 shows another example that has two power domains, a switched power
domain *PD_SW* and an always-on domain *PD_TOP*.

```
create_power_domain PD_TOP -include_scope
# PD_TOP includes all elements at top and
# its children.
create_power_domain PD_SW \
  -elements {U2/U3 U2/U4 U2/RR}
# However, these elements are in PD_SW domain.
```

Fig. 8.3 Another example
of power domains

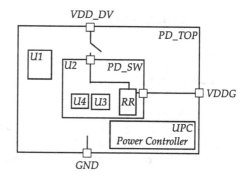

Fig. 8.4 Power and ground
ports of power domains

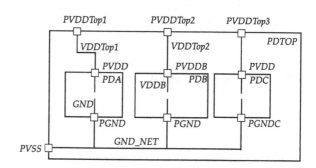

8.3   Creating Supply Ports

For each power domain that is defined, supply and ground ports must be specified
(Fig. 8.4).

```
# First supply ports:
create_supply_port PVDDTop1 -domain PDTOP
create_supply_port PVDDTop2 -domain PDTOP
create_supply_port PVDDTop3 -domain PDTOP
create_supply_port PVDD -domain PDA
create_supply_port PVDDB -domain PDB
create_supply_port PVDD -domain PDC
# Then ground ports:
create_supply_port PVSS -domain PDTOP
create_supply_port PGND -domain PDA
create_supply_port PGND -domain PDB
create_supply_port PGNDC -domain PDC
```

8.4   Creating Supply Nets

A supply net is used to connect the supply ports created.

```
create_supply_net GND_NET -domain PDTOP
create_supply_net VDDTop1 -domain PDTOP
create_supply_net VDDTop2 -domain PDTOP
create_supply_net VDDTop3 -domain PDTOP
create_supply_net GND -domain PDA
create_supply_net VDDB -domain PDB
```

Once a net is created, the -**reuse** option can be used for any subsequent creations of that net in different domains within the same scope. There is also no need to create the subdomain ports or connect to them. Creating a supply net with -**reuse** in a lower domain automatically creates the port and connects it up in that power domain. An example of this is shown later.

8.5   Connecting Supply Nets

The **connect_supply_net** command is used to connect the supply net to one or more supply ports.

```
connect_supply_net GND_NET \
   -ports {PDA/PGND PDB/PGND PDC/PGNDC PVSS}
connect_supply_net VDDTop1 \
   -ports {PVDDTop1 PDA/PVDD}
connect_supply_net VDDTop2 \
   -ports {PVDDTop2 PDB/PVDDB}
connect_supply_net VDDB \
   -ports PVDDB -domain PDB
```

8.6   Primary Supplies of a Domain

This is specified using the **set_domain_supply_net** command. It specifies one primary power and ground connection for every power domain.

```
set_domain_supply_net PDB \
   -primary_power_net VDDB \
   -primary_ground_net GND
```

All cells within a power domain are implicitly connected to the primary power and ground nets of the power domain.

8.7   Creating a Power Switch

The **create_power_switch** command creates a power switch in the specified domain. A power switch has an input supply port and an output supply port. It also has a control port (Fig. 8.5).

The **create_power_switch** command also contains the specification of what control signal causes the switch to turn on.

```
create_power_switch PD_TOP_SW \
  -domain PD_TOP \
  -output_supply_port {SWOUT VDDA_SW} \
  -input_supply_port {SWIN VDDA} \
  -control_port {SWCTL CTRL} \
  -on_state {SW_ON VDDA !CTRL} \
  -off_state {SW_OFF CTRL}
```

In the port specification, the first argument is the port name and the second argument is the net name. The following power switch specification contains a sleep acknowledge port as well, as depicted in Fig. 8.6.

```
create_power_switch PD_GPU_SW \
  -domain PD_GPU \
  -input_supply_port {VDDI VDD_DV} \
  -output_supply_port {VDDO VDD_DV_SW} \
  -control_port {SLEEPI PWRCTL_SLEEP} \
  -control_sense high \
  -ack_port {SLEEPO PWRCTL_ACK} \
  -ack_delay {SLEEPO 1} \
  -on_state {on_state VDD_DV !PWRCTL_SLEEP}
  -off_state {off_state PWRCTL_SLEEP}
```

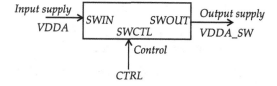

Fig. 8.5 A power switch and its ports

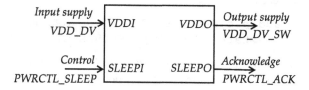

Fig. 8.6 A power switch with acknowledge

8.8   Mapping the Power Switch

This command specifies which cell to use from a technology library for a power switch.

```
map_power_switch PD_GPU_SW \
  -domain PD_GPU \
  -lib_cells PMK/HEADBUF16
```

8.9   State to Supply Port

The **add_port_state** command adds state information to a supply port. It specifies the list of possible states of a port. Each state is specified as a state name and a voltage level. The state name is used to define all possible operating states later. The voltage level can be a single value, two values (minimum and maximum) or a set of three values (minimum, nominal, and maximum) or **off**.

```
add_port_state PD_GPU_SW/VDDO \
  -state {SH_PD_GPU 0.99} \
  -state {SL_PD_GPU 0.79 0.81} \
  -state {OFF_STATE off}

add_port_state PD_TOP_SW/VDDA_SW \
  -state {S_PD_TOP 0.9 1.0 1.1} \
  -state {OFF_STATE off}
```

8.10   Power State Table

A power state table is used to define the legal combination of states that can exist in a design. A power state table captures all possible operating modes of the design in terms of power supply levels.

```
create_pst PD_GPU_PST \
  -supplies {PN1 PN2 SOC/OTC/PN3 FSW/PN4}
```

The above command creates a power state table called *PD_GPU_PST* with the list of four ports.

The **add_pst_state** command defines the combination of port or net state values for each power state.

```
add_pst_state PST0 -pst PD_GPU_PST \
  -state {S0p8 S1p0 S0p9 S1p1}
```

The above command defines the state *PST0* in the power state table *PD_GPU_PST*. The order of entries in the -**state** option is the same as the list of states in the **create_pst** command, which means that *S0p8* is a value of the state of port *PN1*, *S1p0* is the value

Table 8.1 Power state table

State	VDDG	VDD_ON	VDD_SW
PST1	0.8	1.0	1.0
PST2	0.8	1.2	1.2
PST3	0.8	1.0	off
PST4	0.8	1.2	off

of the state of port *PN2*, *S0p9* is the value of state of port *SOC/OTC/PN3* and *S1p1* is the value of state of port *FSW/PN4*. Here is another example.

```
create_pst PST_Y \
  -supplies {VDD GPRS/VDDG1 INST/VDDI}
add_pst_state PST_ST1 \
  -pst PST_Y -state {S0 GPRS_S1 INST_S0}
add_pst_state PST_ST2 \
  -pst PST_Y -state {S0S0S0}
add_pst_state PST_ST3 \
  -pst PST_Y -state {S0S0S1}
add_pst_state PST_ST4 \
  -pst PST_Y -state {S0 off off}
```

The states show the valid power states and the voltage assignments for each state.

Another example of the state table is shown in Table 8.1. The description below shows how it is mapped to a power state table using UPF.

```
# Define the states of the ports first:
add_port_state VDDG \
  -state {S0p8 0.8}
add_port_state VDD_ON \
  -state {S1p0 1.0} \
  -state {S1p2 1.2}
add_port_state VDD_SW \
  -state {S1p0 1.0} \
  -state {S1p2 1.2} \
  -state {sw_off off}

create_pst PST_Z \
  -supplies {VDDG VDD_ON VDD_SW}
add_pst_state PST1 -pst PST_Z \
  -state {S0p8 S1p0 S1p0}
add_pst_state PST2 -pst PST_Z \
  -state {S0p8 S1p2 S1p2}
add_pst_state PST3 -pst PST_Z \
  -state {S0p8 S1p0 sw_off}
add_pst_state PST4 -pst PST_Z \
  -state {S0p8 S1p2 sw_off}
```

8.11   Level Shifter Specification

The **set_level_shifter** command is used to specify the strategy for inserting level shifters. Level shifters are inserted on all nets that have their source and destination at different voltages (which have their source and destination in different power domains).

```
set_level_shifter strategy_name \
  -domain_name domain_name \
  [-element port_pin_list] \
  [-applies_to inputs | outputs | both] \
  [-threshold float] \
  [-rule low_to_high | high_to_low | both ] \
  [-location self | parent | fanout | automatic ] \
  [-non_shift]
```

The -**element** option specifies the list of pins and ports in the domain to which the level shifter strategy is being applied. The -**threshold** option specifies the difference in voltage before a level shifter is inserted. The **low_to_high** value for the -**rule** option specifies to insert level shifters only when going from low to high voltage. The **both** value implies adding level shifters either when going from low to high voltage or when going from high to low voltage. The -**location** option specifies where to place the level shifters. The **self** value indicates to put the level shifter inside the domain whose pins are being level shifted. The **parent** value specifies placing the level shifter in the parent domain. The **fanout** value specifies the placement of the level shifter within all sink (or fanout) domains of the port or pin. The **automatic** value specifies that the tool is free to choose any location. Here is an example of a level shifter strategy that applies to the inputs of a power domain.

```
set_level_shifter LS_INPUTS \
  -domain PD_SHUTDOWN \
  -applies_to inputs \
  -rule low_to_high \
  -location self
```

Here is an example that describes a level shifter strategy for outputs.

```
set_level_shifter LS_OUTPUTS \
  -domain PD_SHUTDOWN \
  -applies_to outputs \
  -rule high_to_low \
  -location parent
```

Figure 8.7 shows where the level shifters are placed based on whether they are inside the power domain or in the parent domain.

Fig. 8.7 Level shifter placement

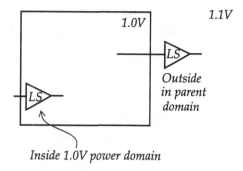

Outside in parent domain

Inside 1.0V power domain

8.12 Isolation Strategy

The **set_isolation** command is used to define an isolation strategy for a power domain. Here is the syntax.

```
set_isolation isolation_strategy_name \
  -domain power_domain \
  [-isolation_power_net isolation_power_net] \
  [-isolation_ground_net isolation_ground_net] \
  [-clamp_value {0 | 1 | latch} \
  [-applies_to {inputs | outputs | both}] \
  [-elements objects] \
  [-no_isolation]
```

At least one of **-isolation_power_net** or **-isolation_ground_net** must be specified. These nets specify the supply nets to be used for isolation logic. The default values for these options are the primary power and primary ground respectively. The **-no_isolation** options applies to the list of elements specified with the **-elements** option that are not to be isolated. When the **-elements** option is not present, it applies to all cells in the power domain. The **-clamp_value** option specifies the value to which the isolation input or output are clamped. The **clamp_value** of **latch** causes the value of the non-isolated port to be latched when the isolation signal becomes active. The **-applies_to** option specifies the parts of the power domain that are to be isolated.

Every **set_isolation** command must have a corresponding **set_isolation_control** command unless the **-no_isolation** option is used. The **set_isolation_control** command contains the specification of the isolation control signal. Here is the syntax.

```
set_isolation_control isolation_strategy_name \
  -domain power_domain \
  -isolation_signal isolation_signal \
  [-isolation_sense {low | high}] \
  [-location {self | parent}]
```

The **isolation_signal** can only be a net; cannot be a port or a pin. The **isolation_sense** specifies the logic state of isolation cells in isolation mode. The **location** value of **self** specifies that the isolation cell be placed within the current hierarchy, and the value of **parent** specifies that the isolation cell is placed in the parent module. Here is an example of the isolation strategy definition and the corresponding isolation control specification.

```
set_isolation ISO_OUTPUT \
  -domain PD_SHUTDOWN \
  -isolation_power_net VDDG \
  -isolation_ground_net VSS \
  -clamp_value 0 \
  -applies_to outputs

set_isolation_control ISO_OUTPUT \
  -domain PD_SHUTDOWN \
  -isolation_signal ISOLATE_CTRL \
  -isolation_sense low \
  -location parent
```

8.13   Retention Strategy

The **set_retention** command specifies which registers in the power domain are to be implemented as retention registers and identifies the save and restore signals. All the registers specified with the **-elements** option are given the retention capabilities. Here is the syntax.

```
set_retention retention_strategy_name \
  -domain domain_name \
  -retention_power_net retention_power_net \
  -retention_ground_net retention_ground_net \
  [-elements objects]
```

At least one of the power net or ground net must be specified. The list of objects can include an instance of a block as well. The retention power and ground nets are connected to the save and restore logic and shadow registers. If the **elements** field is not specified, it is equivalent to using the element list used to define the power domain.

Every retention strategy must have a corresponding **set_retention_control** command. This command allows the specification of the retention control signal and its sense. Its syntax is of the form:

```
set_retention_control ret_strategy_name \
  -domain domain_name \
  -save_signal [save_signal {high | low}] \
  -restore_signal [restore_signal {high | low}]
```

Fig. 8.8 Power controller
with retention signals

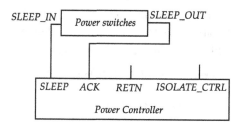

The **save_signal** specifies the net (cannot be a port or a pin) used to save the data in the shadow registers and the logic state of the save signal that causes this to happen. The **restore_signal** is specified in a similar manner. Here are some examples.

```
set_retention UP_RET_POLICY \
  -domain PD_UP \
  -retention_power_net VDDG \
  -retention_ground_net VSS

set_retention_control UP_RET_POLICY \
  -domain PD_UP \
  -save_signal {NSAV low} \
  -restore_signal {RETN high}

set_retention MULT_RET \
  -domain PD_A \
  -retention_power_net VDDB \
  -retention_ground_net VSS
```

A power controller typically generates the signals required for state retention. Figure 8.8 shows an example.

8.14   Mapping the Retention Registers

The **map_retention_cell** command provides a mechanism to specify what cells to use for retention registers. Here is the syntax.

```
map_retention_cell ret_strategy_name \
  -domain domain_name \
  [-lib_cells lib_cells] \
  [-lib_cell_type lib_cell_type] \
  [-lib_model_name lib_cells_name] \
  [-elements objects]
```

The **-lib_cells** provides a list of target library cells that are used for retention mapping. The **-lib_cell_type** specifies the attribute used to identify the cells with retention type that are to be used. The **-lib_model_name** specifies the name of the library cell for this strategy. The **-elements** option specifies the register elements to which the mapping command applies. Here is an example.

```
map_retention_cell MULT_RET \
  -domain PD_A \
  -lib_cells RSDFF_X8T40
```

8.15   Mychip Example

Figure 8.9 shows an example of a design that is used to describe UPF. The top-level *MYCHIP* is always-on at 1.0 V. Block *CPU* is at 0.9 V but is always-on. Block *DSP* is at 1.1 V or at 0.9 V and is switched between these two supplies. Block *COP* is 1.0 V but can be switched off. It has retention registers in it, so that its state can be saved upon shutdown. The power controller is a module at the top level that provides the various control signals for handling the power switching.

The example design has four power domains:

1. *PD_MYCHIP*: 1.0 V, always-on, associated with logic in top-level *MYCHIP*.
2. *PD_CPU*: 0.9 V, always-on, associated with logic in module instance *U_CPU*.
3. *PD_COP*: 1.1 V, shutdown, associated with logic in module instance *U_COP*.
4. *PD_DSP*: 1.1 V or 0.9 V, externally switched, associated with logic in module instance *U_DSP*.

Table 8.2 describes the power mode table, that is, the various power operating modes of the design.

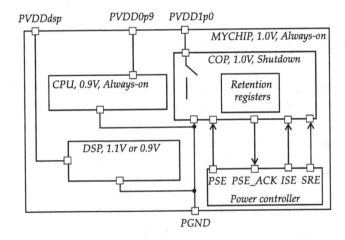

Fig. 8.9   An example with multiple power domains

Table 8.2 Power operating modes

Power mode	PD_MYCHIP	PD_CPU	PD_COP	PD_DSP
PM1	1.0 V	0.9 V	1.0 V	1.1 V
PM2	1.0 V	0.9 V	Off	1.1 V
PM3	1.0 V	0.9 V	1.0 V	0.9 V
PM4	1.0 V	0.9 V	Off	0.9 V

Table 8.3 Signals from power controller

Power domain	Power switch rule	Isolation rule	State retention rule
PD_COP	U_PC.PSE (to switch) U_PC.PSE_ACK (from switch)	U_PC.ISE	U_PC.SRE

The power controller block generates the necessary control signals to control the power domains. These are depicted in Table 8.3.

Here is an example UPF for the *MYCHIP* example.

```
# File: mychip.upf
#
# Set scope to top-level:
set_scope

# Declare power domains:
create_power_domain PD_MYCHIP -include_scope
create_power_domain PD_CPU -elements {U_CPU}
create_power_domain PD_DSP -elements {U_DSP}
create_power_domain PD_COP -elements {U_COP}

# Create power nets at top:
create_supply_net VDD1p0 -domain PD_MYCHIP -reuse
create_supply_net VDDdsp -domain PD_MYCHIP -reuse
create_supply_net VDD0p9 -domain PD_MYCHIP -reuse
create_supply_net GND -domain PD_MYCHIP -reuse

# Create power nets in PD_CPU:
create_supply_net VDD0p9 -domain PD_CPU
create_supply_net GND -domain PD_CPU

# Create power nets in PD_DSP:
create_supply_net VDDdsp -domain PD_DSP
create_supply_net GND -domain PD_DSP

# Create power nets in PD_COP:
create_supply_net VDD1p0 -domain PD_COP
create_supply_net VDD1p0_SW -domain PD_COP
create_supply_net GND -domain PD_COP
```

```
# Create the power ports at top:
create_supply_port PVDD1p0 -domain PD_MYCHIP
create_supply_port PVDD0p9 -domain PD_MYCHIP
create_supply_port PVDDdsp -domain PD_MYCHIP
create_supply_port PGND -domain PD_MYCHIP

# Create power ports in PD_CPU:
create_supply_port PVDD0p9 -domain PD_CPU
create_supply_port PGND -domain PD_CPU

# Create power ports in PD_DSP:
create_supply_port PVDDdsp -domain PD_DSP
create_supply_port PGND -domain PD_DSP

# Create power ports in PD_COP:
create_supply_port PVDD1p0 -domain PD_COP
create_supply_port PGND -domain PD_COP

# Connect top power ports and nets:
connect_supply_net VDD1p0 -ports PVDD1p0
connect_supply_net VDD0p9 -ports PVDD0p9
connect_supply_net VDDdsp -ports PVDDdsp
connect_supply_net GND -ports PGND

# Connect top to PD_CPU:
connect_supply_net VDD0p9 -ports {U_CPU/PVDD0p9}
connect_supply_net GND -ports {U_CPU/PGND}

# Connect top to PD_DSP:
connect_supply_net VDDdsp -ports {U_DSP/PVDDdsp}
connect_supply_net GND -ports {U_DSP/PGND}

# Connect top to PD_COP:
connect_supply_net VDD1p0 -ports {U_COP/PVDD1p0}
connect_supply_net GND -ports {U_COP/PGND}

# Connect inside PD_CPU:
connect_supply_net VDD0p9 \
  -ports PVDD0p9 -domain PD_CPU
connect_supply_net GND -ports PGND -domain PD_CPU

# Connect inside PD_DSP:
connect_supply_net VDDdsp \
  -ports PVDDdsp -domain PD_DSP
connect_supply_net GND -ports PGND -domain PD_DSP

# Connect inside PD_COP:
connect_supply_net VDD1p0 \
  -ports PVDD1p0 -domain PD_COP
connect_supply_net GND -ports PGND -domain PD_COP
```

```
# Specify primary power nets:
set_domain_supply_net PD_MYCHIP \
  -primary_power_net VDD1p0 \
  -primary_ground_net GND
set_domain_supply_net PD_CPU \
  -primary_power_net VDD0p9 \
  -primary_ground_net GND
set_domain_supply_net PD_DSP \
  -primary_power_net VDDdsp \
  -primary_ground_net GND
set_domain_supply_net PD_COP \
  -primary_power_net VDD1p0_SW \
  -primary_ground_net GND

# Define isolation strategy and control for PD_COP:
set_isolation PD_COP_ISO \
  -domain PD_COP \
  -isolation_power_net VDD1p0 \
  -isolation_ground_net GND \
  -applies_to outputs \
  -clamp_value 0

set_isolation_control PD_COP_ISO \
  -domain PD_COP \
  -isolation_signal U_PC/ISE \
  -isolation_sense low \
  -location self

# Define level shifter strategy and
# control for PD_CPU:
set_level_shifter FROM_PD_CPU_LST \
  -domain PD_CPU \
  -applies_to outputs \
  -rule low_to_high \
  -location parent

set_level_shifter TO_PD_CPU_LST \
  -domain PD_CPU \
  -applies_to inputs \
  -rule high_to_low \
  -location self

# Define level shifter strategy and
# control for PD_DSP:
set_level_shifter FROM_PD_DSP_LST \
  -domain PD_DSP \
  -applies_to_outputs \
```

```
  -rule low_to_high \
  -location parent

set_level_shifter TO_PD_DSP_LST \
  -domain PD_DSP \
  -applies_to inputs \
  -rule high_to_low \
  -location self

# Declare the switches for PD_COP:
create_power_switch PD_COP_SW \
  -domain PD_COP \
  -input_supply_port {VDDG VDD1p0} \
  -output_supply_port {VDD VDD1p0_SW} \
  -control_port {SLEEP U_PC/PSE} \
  -ack_port {SLEEPOUT U_PC/PSE_ACK} \
  -ack_delay {SLEEPOUT 10} \
  -on_state {SW_on VDDG !SLEEP} \
  -off_state {SW_off SLEEP}

# Specify the switch type:
map_power_switch PF_COP_SW \
  -domain PD_COP \
  -lib_cells HEADBUF_T50

# Specify isolation cell type:
map_isolation_cell PD_COP_ISO \
  -domain PD_COP \
  -lib_cells {O2ISO_T50 A2ISO_T50}

# Specify retention strategy:
set_retention PD_COP_RET \
  -domain PD_COP \
  -retention_power_net VDD1p0 \
  -elements {U_COP/reg1 U_COP/pc U_COP/int_state}

set_retention_control PD_COP_RET \
  -domain PD_COP \
  -save_signal {U_PC/SRE high} \
  -restore_signal {U_PC/SRE low}

# Add port state:
add_port_state PVDD1p0 \
  -state {S1p0 1.0}
add_port_state PVDD0p9 \
  -state {S0p9 0.9}
add_port_state PVDDdsp \
  -state {SH1p1 1.1}\
```

```
  -state {SL1p1 0.9}
add_port_state PGND \
  -state {default 0}
add_port_state PD_COP_SW/VDD \
  -state {SW_on 1.0} \
  -state {SW_off off}

# Create power states and state table:
create_pst MYCHIP_pst -supplies \
  {VDD1p0 VDD0p9 VDDdsp PD_COP_SW/VDD}
add_pst_state PM1 -pst MYCHIP_pst -state \
  {S1p0 S0p9 SH1p1 SW_on}
add_pst_state PM2 -pst MYCHIP_pst -state \
  {S1p0 S0p9 SH1p1 SW_off}
add_pst_state PM3 -pst MYCHIP_pst -state \
  {S1p0 S0p9 SL1p1 SW_on}
add_pst_state PM4 -pst MYCHIP_pst -state \
  {S1p0 S0p9 SL1p1 SW_off}
```

The complete UPF syntax is described in Appendix B.

Chapter 9
CPF Power Specification

This chapter describes the Common Power Format (CPF), which is yet another power specification language.

9.1 Introduction

The Common Power Format is a standard promoted by the Low Power Coalition at Si2.[1] CPF 1.1 was released in 2008. CPF 2.0 was released in 2011 [CPF11].

CPF is a TCL-based language that operates on specification objects and the design objects. A design object is a module, instance, net, pin or port as they appear in the RTL.

The language can be used to express power design intent in terms of:

1. *Power domains*: From physical, logical and analysis point of view
2. *Power logic*: For example, level shifters and isolation logic
3. *Power mode*: State mode table and transitions

The set of commands in CPF can be broken down into the following categories:

1. Library commands
2. Hierarchical support commands
3. General purpose commands
4. Power mode commands
5. Design and implementation constraints
6. Macro support commands
7. Version and verification support commands

---

[1] www.si2.org

R. Chadha and J. Bhasker, *An ASIC Low Power Primer: Analysis, Techniques and Specification*, DOI 10.1007/978-1-4614-4271-4_9,
© Springer Science+Business Media New York 2013

9.2   Library Commands

9.2.1   Define Always-on Cell

The command identifies the cells that are to be used as always-on cells. These cells have an additional set of power and/or ground pins that can stay powered *on* even if the power domain that they are instantiated in is shut down.

```
define_always_on_cell \
  -cells {AOBUF2 AOBUF3 AOBUF4} \
  -ground VSSG \
  -ground_switchable VSS \
  -power VDDG \
  -power_switchable VDD
```

The cells *AOBUF2*, *AOBUF3* and *AOBUF4* are the always-on cells. The -**ground** and -**power** options specify the pins that are always powered. The -**ground_switchable** and the -**power_switchable** options specify the pin names of the power domain that is shut down.

9.2.2   Define Global Cell

The command identifies cells that have more than one set of power and ground pins and retain their functional behavior even if the local power and local ground are switched off.

```
define_global_cell \
  -cells NOR2X8_AON \
  -local_power VDD \
  -local_ground VSS \
  -global_power VDDG \
  -global_ground VSSG
```

The cell *NOR2X8_AON* continues to function even when the local power *VDD* is switched off.

9.2.3   Define Isolation Cell

The command identifies the cells that are to be used as isolation cells.

```
define_isolation_cell \
  -cells ISO* \
```

```
-enable EN \
-valid_location from \
-always_on_pins {Y Z}
```

The enable pin of the *ISO** cells is *EN*, and the valid location of the isolation cell is in the source power domain. The **-always_on_pins** specifies a list of pins that are related to the non-switchable power and non-switchable ground pins of the cells.

9.2.4   Define Level Shifter Cell

The command identifies the cells that are to be used as level shifter cells.

```
define_level_shifter_cell \
  -cells LVL* \
  -input_voltage_range 1.2 \
  -output_voltage_range 0.8 \
  -direction down \
  -valid_location to
```

The **-cells** specifies the level shifter cells to be *LVL**. The **-input_voltage_range** specifies the input voltage range that this level shifter can handle. The **-output_voltage_range** specifies the output voltage that the level shifter can handle. The **-direction** specifies that it can be used to go from a higher voltage domain to a lower voltage domain. The **-valid_location** option specifies that the level shifter should be placed in the destination power domain.

Here is another example.

```
define_level_shifter_cell \
  -cells LSLH* \
  -direction down \
  -valid_location either \
  -input_power_pin VL \
  -output_power_pin VH \
  -ground GND
```

The defined cells are to be used when going from a higher voltage domain to a lower voltage domain. The cells can be placed in either the source power domain or the destination power domain. The **-input_power_pin** is the name of the power pin in the LEF[2] file that must be connected to the voltage of the source power domain. The **-output_power_pin** specifies the name of the power pin in the LEF file that must be connected to the destination power domain. The **-ground** option identifies the name of the ground pin in the LEF file.

---

[2] A LEF file contains the IO pin descriptions of the physical layout of the cell. See [BHA06].

9.2.5   Define Open Source Input Pin

This command specifies a list of cells that contain open *source* input pins. These input pins must be isolated when the power to the driver is *on*, but the power of the cell that the input pin is connected to is *off*.[3]

```
define_open_source_input_pin \
  -cells OSL \
  -pin A0
```

9.2.6   Define Pad Cell

This command identifies pad cells in the library.

```
define_pad_cell \
  -cells PDBX2 \
  -pad_pins PAD
```

9.2.7   Define Power Clamp Cell

This command specifies a list of diode cells for power clamp control.

```
define_power_clamp_cell⁴ \
  -cells PCMP \
  -data D \
  -power VDD \
  -ground VSS
```

The data pin is *D*, the power pin is *VDD* and the ground pin is *VSS*.

9.2.8   Define Power Clamp Pins

This command specifies a list of cells that are either power or ground clamp cells.

```
define_power_clamp_pins \
  -cells PVDDCL \
  -data_pins IN
```

The **data_pins** option specifies a list of cell input pins that have clamp diodes.

---

[3] It is a good design practice to isolate *all* input pins when the power of the cell is *off* but the power to the driver is *on*.

[4] The **define_power_clamp_cell** has been superseded by the **define_power_clamp_pins** command.

9.2.9   Define Power Switch Cell

The command identifies the cells that are to be used as power switch cells.

```
define_power_switch_cell \
  -cells {GEAD2 GEAD3} \
  -ground VSS \
  -power_switchable VDD \
  -power VDDG \
  -stage_1_enable SLEEP \
  -stage_1_output SLEEPOUT \
  -type header
```

The **-ground** option identifies the input ground pin in the LEF file. The **-power_switchable** option identifies the output power pin in the LEF file that must be connected to a switchable power net. The **-power** option specifies the input power pin in the LEF file. The **-stage_1_enable** specifies the expression that must be true for the switch to be turned on. The **-stage_1_output** specifies the output pin of the switch. The **-type** specifies whether the cell is a header cell or a footer cell.

Here is another example.

```
define_power_switch_cell \
  -cells PSCELL \
  -stage_1_enable !SL1IN \
  -stage_1_output SL1OUT \
  -stage_2_enable SL2IN \
  -stage_2_output !SL2OUT \
  -stage_1_on_resistance 200 \
  -stage_2_on_resistance 10\
  -type footer \
  -enable_pin_bias 0:0.2
```

The *stage1* transistor is turned on when *SL1IN* is low. The *stage1* output *SL1OUT* is a buffered output of input *SL1IN*. The *stage2* transistor is turned on when *SL2IN* is high. And the *stage2* output *SL2OUT* is an inverted value of input *SL2IN*. The **-enable_pin_bias** option specifies that the enable pin of the power switch cell can be driven up to $VDD + 0.2$ V.

9.2.10   Define Related Power Pin

This command specifies the relationship between the power pins and data pins for cells that have more than one set of power and ground pins.

```
define_related_power_pins \
  -data_pins pin_list \
```

```
-cells cell_list \
-power VDDG \
-ground VSSG
```

The **-data_pins** specify a list of input and output data pins. The **-cells** option specifies the cells for which the relationship between the power and data pins are defined.

9.2.11   Define State Retention Cell

The command specifies the cells that are to be used as state retention cells.

```
define_state_retention_cells \
  -cells "DFRR" \
  -restore_function RETN
```

The **-restore_function** option specifies the expression that causes the retention cell to restore the saved value after exiting shutdown mode. The expression must be an always-on expression.

Here is another example.

```
define_state_retention_cell \
  -cells "TSFF" \
  -power VDDC \
  -ground VSS \
  -power_switchable VDD \
  -restore_function "my_restore" \
  -restore_check "!CSAVE" \
  -save_function "!PG"
```

The option **-restore_check** specifies that signal *CSAVE* must be held low to save the state of the sequential cell. The **-save_function** option specifies that signal *PG* must be held low, then the state will be saved. The **-ground** option specifies the ground pin of the cell. The **-power** option specifies the power pin of the cell; this is also the power pin that is always *on* when the cell is in shutdown mode. The **-power_switchable** option specifies the power pin of the cell that is turned off during shutdown mode.

9.3   Power Mode Commands

9.3.1   Create Mode

This command defines a mode of the design.

```
create_mode \
  -name FUNC3 \
  -condition {PD_GPU@OFFV}
```

The design is in *FUNC3* mode when the power domain *PD_GPU* is shut down. *OFFV* is a condition defined as:

```
create_nominal_condition -name OFFV -voltage 0
```

9.3.2 Create Power Mode

This command defines a power mode. One of the power modes must be specified as the default power mode.

```
        PDA     PDB     PDC
PM_A    on      1.0V    0.9V
PM_B    on      off     0.9V
PM_C    on      1.0V    1.0V
```

```
create_nominal_condition -name V0p9 -voltage 0.9
create_nominal_condition -name V1p0 -voltage 1.0

create_power_mode \
  -name PM_A \
  -default \
  -domain_conditions {PDA@on PDB@V1p0 PDC@V0p9}
create_power_mode -name PM_B \
  -domain_conditions {PDA@on PDB@off PDC@V0p9}
create_power_mode -name PM_C \
  -domain_conditions {PDA@on PDB@V1p0 PDC@V1p0}
```

The **-default** option specifies that this mode corresponds to the initial state of the design. The **-domain_conditions** option specify the power mode conditions which is of the form *domain_name@nominal_condition_name*.

9.3.3 Specify Power Mode Transition

The command **create_mode_transition** specifies how the transition between two power modes is controlled and the time it takes for each power domain to complete the transition. Each mode transition involves a change in one or more power domains from the previous condition to the new condition. The start and end conditions can be used by a verification tool. Similarly the transition time can be used to simulate the latency of each mode transition.

Its syntax is of the form:

```
create_mode_transition \
  -name string \
  -from_mode power_mode \
  -to_mode power_mode \
  { -start_condition expression \
    [-end_condition expression] \
    [-cycles [integer:]integer \
      -clock_pin clock_pin \
    | -latency [float:]float] \
  }
```

Here are two examples.

```
create_mode_transition \
  -name A2B \
  -from_mode PM_A \
  -to_mode PM_B \
  -start_condition {!UPC/INIT} \
  -clock_pin {UPC/PCLK} \
  -cycles 100

create_mode_transition \
  -name CMP \
  -from PMA \
  -to PMB \
  -latency 12 \
  -start_condition RSTB
```

In the second example, the transition between modes *PMA* to *PMB* is controlled by the signal *RSTB* and the time it takes to transition from one mode to the other is specified using the **-latency** option.

9.3.4 Set Power Mode Control Group

The command groups a list of power domains into power mode control groups. This command along with the **end_power_mode_control_group** command groups a set of CPF commands that apply to the group only. Only the following CPF commands are allowed to be specified in such a group: **create_analysis_view**, **create_mode_transition**, **create_power_mode**, and **update_power_mode**.

```
set_power_mode_control_group \
  -name PMCG1 \
  -domains {PD1 PD2}
```

The **-domains** specifies the list of power domains within the specified group.

9.3.5   End Power Mode Control Group

This command is used in conjunction with the **set_power_mode_control_group** and it associates a set of CPF commands with a group.

```
end_power_mode_control_group
```

9.4   Design and Implementation Constraints

9.4.1   Create Analysis View

The command creates an analysis view and associates a list of operating corners with a given mode.

```
create_analysis_view \
  -name SLOW_VIEW \
  -mode PM1 \
  -domain_corners {PDcore@120v_fast \
                   PDalu@120v_slow}
```

The **-name** option specifies the name of the analysis view. The **-mode** option specifies a mode that must have been previously created using **create_power_mode**.

9.4.2   Create Bias Net

The command specifies a bias net that can be used as a power supply to either forward or backward body bias a transistor.

```
create_bias_net \
  -net VGG_BIAS \
  -driver VGGB
```

The net *VGG_BIAS* is declared as a bias net driven by pin *VGGB*.

9.4.3   Create Global Connection

This command specifies how a global net is connected to the specified pins or ports.

```
create_global_connection \
  -net N0 \
  -pins {P1 P2 P3}
```

In above, the global connection applies to the specified pins of the entire design. One can use the **-domain** or **-instances** option to qualify the global connection of the pins.

9.4.4 Create Ground Net

The following command specifies the ground nets. Here is the syntax followed by an example.

```
create_ground_nets \
  -nets list_of_nets \
  [-voltage {float | voltage_range}] \
  [-external_shutoff_condition expression | \
   -internal] \
  [-user_attributes string_list] \
  [-peak_ir_drop_limit float] \
  [-average_ir_drop_limit float]

create_ground_nets \
  -nets VSS \
  -voltage 0.0
```

The **-voltage** option identifies the voltage that is applied to the net *VSS*.

9.4.5 Create Isolation Rule

The command specifies which nets are to be isolated. It specifies a list of nets that need isolation when the power domain for the source or destination are switched off. In most cases, isolation is needed when the source (or driver) is switched off but the destination is still on. In most cases, an isolation cell is not needed when the source is on and the destinations are off.

```
create_isolation_rule \
  -name ISO_RULE1 \
  -from PD_BLK1 \
  -isolation_condition {!UPC/PAU} \
  -isolation_output low \
  -isolation_target from
```

The name of the isolation rule is *ISO_RULE1*. The **-from** option specifies a power domain from which all nets need to be isolated. The -isolation_condition specifies the condition under which these nets need to be isolated. The **-isolation_output** specifies the output value of the isolation cells when the isolation condition is true. The **-isolation_target** specifies that the rule applies when the drivers (can also be **-to** to indicate the sink of a net) of the nets are turned off.

In the following example,

```
create_isolation_rule \
  -name PD_ALU_ISO \
  -from PD_ALU \
  -to PD_TOP \
  . . .
```

the isolation rule *PD_ALU_ISO* applies to all nets that have their drivers in the *PD_ALU* power domain and the destinations in the *PD_TOP* power domain.

9.4.6   Create Level Shifter Rule

The command defines a rule for adding level shifters.

```
create_level_shifter_rule \
  -name LS0 \
  -to PD_COP \
  -exclude U0/Z
```

The name of the rule is *LS0* and all nets that go to the power domain *PD_COP* are to be level shifted. The **-exclude** option specifies the pins whose nets are not to be level shifted.

9.4.7   Create Nominal Condition

The command creates an operating condition with the specified voltage.

```
create_nominal_condition \
  -name NCONDA \
  -voltage 1.1 \
  -ground_voltage 0.0 \
  -state off \
  -pmos_bias_voltage 0.2 \
  -nmos_bias_voltage 0.2
```

The **-voltage** option specifies the power supply voltage. The **-ground_voltage** option specifies the ground supply voltage. The **-state** option specifies the state of a power domain when this nominal condition is used. The **-pmos_bias_voltage** specifies the body bias voltage for the p-type transistors. The **-nmos_bias_voltage** specifies the body bias voltage for the n-type transistors.

9.4.8   Create Operating Corner

The command defines an operating corner and links it with a library set which has been defined earlier with the **define_library_set** command.

```
create_operating_corner \
  -name FAST1P2V \
  -voltage 1.2 \
  -temperature 0C \
  -library_set FF_LIBS
```

The name of the corner is *FAST1P2V*. It operates at a voltage of 1.2 V, temperature of 0C, and is associated with a library set with the name *FF_LIBS*.

9.4.9   Create Pad Rule

The command specifies the mapping of pad instances to top-level power domains.

```
create_pad_rule \
  -name MYCHIP_PAD_RULE \
  -mapping {{CIN PD_MYCHIP} {DEFAULT PD_IO}}
```

The pad rule *MYCHIP_PAD_RULE* specifies that the pad pin *CIN* belongs to the *PD_MYCHIP* power domain and all other pad pins belong to the *PD_IO* power domain.

9.4.10   Create Power Domain

The command defines a power domain and specifies all the instances that belong to that power domain.

```
create_power_domain \
  -name PD_top \
  -default
```

The **-default** option specifies that this is the default power domain. Any instances that are not associated with any power domain belong to the default power domain.
In the following example,

```
create_power_domain \
  -name PD_A \
  -instances U_BLKA
```

instance *U_BLKA*, and all its sub-instances, belong to the *PD_A* power domain.

Here is another example.

```
create_power_domain \
  -name PD_B \
  -instances {U_BLKB UBLKC} \
  -shutoff_condition U_PC/SWITCH
```

The **-shutoff_condition** option specifies the condition when the power domain *PD_B* is shut down. *SWITCH* is a port of the power controller module *U_PC*.

In the following example,

```
create_power_domain \
  -name PD_C \
  -instances U_BLKC \
  -shutoff_condition U_PC/blkc_shutoff \
  -power_up_states low \
  -active_state_conditions nom_cond@foo \
  -boundary_ports {A B C}
```

The **-power_up_states** option specifies the power-up initialization states of the non-state retention cells of the power domain. The **-active_state_conditions** specify the boolean condition under which the power domains are considered on. The **-boundary_ports** option specifies the list of inputs and outputs that are part of the power domain.

9.4.11   Create Power Net

The command defines a list of power nets.

```
create_power_nets \
  -nets {VDD1 VDD3} \
  -voltage 1.2
```

VDD1 and *VDD3* are power nets with a voltage of 1.2 V. Here is another example.

```
create_power_nets \
  -nets VDD2 \
  -voltage 0.8v \
  -external_shutoff_condition WATCH_F1 \
  -average_ir_drop_limit 0.03 \
  -peak_ir_drop_limit 0.05
```

The net *VDD2* is powered by an external power source and the **-external_shutoff_condition** specifies the condition under which the power is turned off on the net. The IR drop[5] limits specify the average and peak IR drop limits allowed on the specified net. Here is another example.

---

[5] Voltage drop in a power mesh.

```
create_power_nets \
  -nets {VDD_core VDD_alu} \
  -internal
```

The -**internal** option specifies that the nets are driven by an on-chip power switch.

9.4.12   Create Power Switch Rule

The command specifies how an on-chip power switch must connect the external power and ground nets to the power and ground nets of the specified power domain.

```
create_power_switch_rule \
  -name PSR_AU \
  -domain PD_AU \
  -external_power_net VDD1
```

The -**external_power_net** option specifies the external power net to which the source pin of the power switch is to be connected (only for a header cell). Here is another example.

```
create_power_switch_rule \
  -name PSR_CORE \
  -domain PD_CORE \
  -external_ground_net VSS1
```

The -**external_ground_net** option specifies the external ground net to which the source pin of the ground switch is to be connected (only for a footer cell).

9.4.13   Create State Retention Rule

The command specifies the registers in the domain that are to be replaced with state retention registers.

```
create_state_retention_rule \
  -name SRR_A \
  -domain PD_A \
  -restore_edge {!PCU/PRF}
```

The state retention rule *SRR_A* applies to all flip-flops in domain *PD_A*. The -**restore_edge** option specifies the condition under which the states are restored. When *PRF* goes high, the flip-flops get into retention mode. When *PRF* goes low, the retention mode ends and the saved data appears on the flip-flop outputs. Here is another example.

```
create_state_retention_rule \
  -name SRR_B \
  -instances {INST1 INST2} \
  -exclude {INST1/FF1 INST2/FF4} \
  -save_edge PRF \
  -save_precondition RST_B \
  -target_type flop
```

The state retention rule *SRR_B* applies only to the specified instances. The **-exclude** option specifies the flip-flops that are to be excluded from retention. The **-save_edge** option specifies the condition that causes the states to be saved. The **-target_type** indicates that only flip-flops are to be replaced (other than **flop**, the options are **latch** and **both**).

The save signal should come from the parent before the power domain is switched off. After the power domain is restored and the restore signal is activated, the register value is restored. The command specifies the list of registers to be mapped to retention registers. If no **-instances** option is present, it applies to all flip-flops present in that power domain.

9.4.14 Define Library Set

The command defines a set of libraries.

```
define_library_set \
  -name LIBSET_A \
  -libraries {stdcell.lib iocell.lib}
```

9.4.15 Identify Always-on Driver

The command specifies a list of pins in the design that must be driven by always-on buffers or always-on inverters.

```
identify_always_on_driver \
  -pins {U0/S1 U1/U5/Z}
```

9.4.16 Identify Power Logic

The command identifies any isolation logic that is instantiated in the design.

```
identify_power_logic \
  -type isolation \
  -instances {U0 U1 U2}
```

9.4.17   Identify Secondary Domain

The command identifies the secondary power domain for the specified instances that have multiple power and ground pins.

```
identify_secondary_domain \
  -secondary_domain VDDA \
  -instances {U0 U1} \
  -domain PDA
```

The instances *U0* and *U1* in the *PDA* domain have multiple power and ground pins.

9.4.18   Specify Equivalent Control Pins

The command specifies a list of pins that are equivalent with a master control pin.

```
set_equivalent_control_pins \
  -master pinA \
  -pins {pinA pinB} \
  -domain PDA
```

The -**domain** option specifies the domain for which the master control pin is part of the shutdown condition. The -**master** option specifies the name of the master pin.

9.4.19   Specify Input Voltage Tolerance

The command specifies the voltage range that can be tolerated at the specified pins without the need for level shifters.

```
set_input_voltage_tolerance \
  -power -0.2:0.2 \
  -ground -0.1
```

The input power voltage can be 0.2 V less or 0.2 V more than the standard operating voltage and the input ground voltage can be 0.1 V less than the standard ground voltage without requiring a level shifter.

9.4.20   Set Power Target

The command specifies the target for the average leakage power and average dynamic power of the design. The power unit is specified using the **set_power_unit** command.

```
set_power_target \
  -leakage 2.11 \
  -dynamic 5.5
```

9.4.21   Set Switching Activity

The command is used to specify the toggle rate and the probability[6] for the specified pins.

```
set_switching_activity \
  -all \
  -probability 0.2 \
  -toggle_rate 1
```

The **-all** option specifies that the switching activity applies to all pins of design. The **-probability** value should be between 0 and 1 and specifies the static probability value is 0.2. The **-toggle_rate** specifies the number of toggles per time unit. If the time unit is a *ns*, then the value of 1 specifies that the pin toggles 1 times per *ns*, which is equivalent to a frequency of 500 MHz.

9.4.22   Update Isolation Rule

The command adds additional implementation information to an isolation rule.

```
update_isolation_rules \
  -names {IR_A IR_B} \
  -location to \
  -cells {ISOH ISOL} \
  -prefix CPF_ISOR
```

The rules must have been defined previously using the **create_isolation_rule** command. The **-location** option specifies that the isolation logic is to be inserted with the instances of the destination power domain. The default is **to** if this option is not specified. The -**prefix** option specifies the prefix to be used when creating the isolation logic. The **-cells** option specifies the names of the library cells that must be used as isolation cells. If this option is not specified, then the one specified with **define_ isolation_cell** is used.

---

[6] Same as **static_probability.**

9.4.23   Update Level Shifter Rule

The command adds additional implementation information to an existing level shifter rule. The command is optional and if not specified, the rule according to the **define_level_shifter_cell** command is applied.

```
update_level_shifter_rules \
  -names {LSR_A LSR_B} \
  -location from \
  -cells CKLS \
  -prefix CPF_LS
```

The rules must have been previously defined with the **create_level_shifter_rule** command. The **from** value for the -**location** option indicates that the level shifters should be placed in the source power domain. The default value is **to**. The -**cells** option specifies the names of the library cells that are to be used as level shifters. And the -**prefix** option specifies the prefix to be used when creating this logic.

9.4.24   Update Nominal Condition

The command associates a library set with the specified nominal operating condition.

```
update_nominal_condition \
  -name high \
  -library_set WCL_120V
```

The library set must have been defined earlier using the **define_library_set** command. The -**name** option specifies the name of the nominal operating condition (defined earlier with **create_nominal_condition**).

9.4.25   Update Power Domain

This command adds additional implementation aspects for a power domain. Here is the syntax.

```
update_power_domain \
  -name domain \
  [-instances instance_list] \
  [-boundary_ports port_list] \
  {-primary_power_net net \
  | -primary_ground_net net \
  | -equivalent_power_nets power_net_list \
```

```
    | -equivalent_ground_nets ground_net_list \
    | -pmos_bias_net net \
    | -nmos_bias_net net \
    | -deep_nwell_net net \
    | -deep_pwell_net net \
    | -user_attributes string_list \
    | -transition_slope [float:]float \
    | -transition_latency {from_nom latency_list} \
    | -transition_cycles {from_nom cycle_list \
                          clock_pin} \
}
```

Here is an example.

```
update_power_domain \
    -name PDcore \
    -primary_power_net VDD_core \
    -primary_ground_net VSS \
    -equivalent_ground_nets {V1 V2} \
    -equivalent_power_nets {P1 P2}
```

The **-name** option specifies the name of the power domain. The **-primary_power_net** specifies the primary power net for all gates in the power domain (declared earlier with **create_power_nets** command). The **-primary_ground_net** specifies the primary ground net for all gates in the power domain (specified earlier using **create_power_nets** command). The **-equivalent*** options specify the set of power nets that are equivalent to the primary power net (and ground net) of the power domain.

9.4.26 Update Power Mode

The command specifies additional constraints for a power mode.

```
update_power_mode \
    -mode PM1 \
    -sdc_files pm1.sdc
```

The **-mode** option specifies the mode to which the constraints apply to. The **-sdc_files** option specifies the list of SDC files that are to be used for the specified mode. Here is another example.

```
update_power_mode \
    -name PM4 \
    -sdc_files pm4.sdc \
    -activity_file top.vcd \
    -activity_file_weight 0.5
```

The **-activity_file** specifies the path to the activity file. The file format could be either in VCD,[7] TCF[8] or SAIF. The **-activity_file_weight** option specifies the relative weight of the activities in the file; this number can be any number between 0 and 100. To estimate total average chip power over all modes, the activity weights are used to adjust the relative weight of each power mode.

Here is the complete list of options for this command.

```
update_power_mode \
  -name mode \
  { -activity_file file \
   -activity_file_weight weight \
  | -sdc_files sdc_file_list \
  | -setup_sdc_files sdc_file_list \
  | -hold_sdc_files sdc_file_list \
  | -peak_ir_drop_limit domain_voltage_list \
  | -average_ir_drop_limit domain_voltage_list \
  | -leakage_power_limit float \
  | -dynamic_power_limit float \
  }
```

9.4.27   Update Power Switch Rule

The command updates additional information to a power switch rule that has already been created earlier (using **create_power_switch_rule**).

```
update_power_switch_rule \
  -name PS1 \
  -enable_condition_1 ENA \
  -enable_condition_2 ENB \
  -prefix PS_ \
  -cells HDMDCOL1X
```

The **-cells** option specifies the name of the library cell that will be used as a power switch cell. It must have been declared earlier with a **define_power_switch_cell** command. The **-enable_condition** specifies the conditions under which the power switch is to be enabled; one for each *stage1* and *stage2*. The **-name** option is the power switch rule that must have been declared earlier using the **create_power_switch** rule. The **-prefix** specifies the prefix to be used when creating the new logic; the default prefix is "CPF_PS_".

---

[7] For VCD, see [BHA06].

[8] Toggle Count Format.

9.4.28 Update State Retention Rule

The command updates the specified state retention rule with additional information. The rule must have been already created using the **create_state_retention_rule** command. Here is the syntax.

```
update_state_retention_rules \
   -names rule_list \
   { -cell_type string \
   | -set_reset_control \
   | -cells cell_list \
     [-use_model \
        -pin_mapping pin_mapping_list \
        [-domain_mapping domain_mapping_list]] \
   }
```

Here is an example.

```
update_state_retention_rules \
   -names SL1 \
   -cells SRFD1S \
   -cell_type SRFD1S
```

The **-names** option specifies the rule that is being updated. The **-cells** option specifies a list of library cells that can be used to map the sequential cells. The **-cell_type** option specifies the class of library cells that can be used to map the flip-flops; these library cells must have been defined earlier using the **define_state_retention_cell** command.

9.5 Hierarchical Support Commands

9.5.1 End Design

The command, used in conjunction with the **set_design** or an **update_design** command, defines the group of CPF commands that are to be applied to the design.

```
end_design MYCHIP
```

9.5.2 Get Parameter

The command returns the value of a parameter. The parameter must have been defined with the **-parameters** option in the **set_design** command.

```
get_parameter BYPASS_CHECK
```

9.5.3 Set Design

The command specifies the name of the design to which the CPF commands that follow apply to.

```
set_design RX \
  -ports {VP1 VP2} \
  -parameters {{A 1} {B 3}}
```

The -ports option specifies a list of virtual ports of the module. These ports do not exist, thus called virtual, but will be needed to specify control signals for the low power logic. The -parameters option specifies a list of parameters and their values.

9.5.4 Set Instance

The command changes the scope to the specified instance.

```
set_instance INST1 \
  -port_mapping {{CTRL CPU/CTRL} \
  {SAVE CPU/SAVE}}
```

The -port_mapping option specifies the mapping of the virtual ports specified in the set_design command to the drivers in the parent design.

```
set_instance INST2 -design TOP
```

The above example specifies a link to the previously loaded CPF model for *TOP* to the instance *INST2*. The following:

```
set_instance
```

simply returns the current scope.

9.5.5 Update Design

This command adds the power intent specified between this command and the end_design command to the specified power design.

```
update_design MYCHIP
  . . . // CPF commands here.
end_design MYCHIP
```

9.6 General Purpose Commands

9.6.1 Find Design Objects

This command searches and returns design objects that match the specified criteria.

```
find_design_objects *ISO \
  -pattern_type cell \
  -hierarchical
```

The above command searches for all cells that end in *ISO* and returns a list of such cells. The -**hierarchical** option causes the search to be done hierarchically downwards from the current scope.

9.6.2 Specify Array Naming Style

The command specifies the naming style used in the netlist for a multi-bit array in RTL.

```
set_array_naming_style \[%d]\
# Will be of form A[0], A[1], A[2], etc.
```

This is also the default naming style.

9.6.3 Specify Hierarchy Separator

The command specifies the hierarchy separator character that is used in the CPF description.

```
set_hierarchy_separator /
```

The default hierarchy separator character is ".".

9.6.4 Specify Power Unit

The command is used to specify the unit of power in the CPF file.

```
set_power_unit mW
# Can be pW, nW, uW or W as well.

set_power_unit
# Returns the current power unit. The default is mW.
```

9.6.5   Specify Register Naming Style

The command specifies the naming convention used to name flip-flops and latches in the netlist starting from an RTL description.

```
set_register_naming_style _flop%s
```

The "_flop" and the bit number is appended to each flip-flop and latch in the corresponding netlist. The default naming style is to append with "_reg%s".

```
set_register_naming_style
# Returns the current setting.
```

9.6.6   Specify Time Unit

The command specifies the time unit for values in the CPF file.

```
set_time_unit ns
# Could also be us and ms. Default is ns.
```

```
set_time_unit
# Returns the current time unit.
```

9.6.7   Specify Include File

The command can be used to include another CPF file or a TCL file.

```
include ../IP/pll.cpf
```

9.7   Macro Support Commands

9.7.1   Specify Macro Model

The command specifies the start of the CPF content for a custom IP.[9] The macro model is treated as a black-box.

```
set_macro_model ram25x3
. . .
end_macro_model ram25x3
```

---

[9] Intellectual Property.

9.7.2 End Macro Model

The command is used in conjunction with the **set_macro_model** command and encapsulates a set of CPF commands that apply to the macro model.

end_macro_model [*macro_cell_name*]

9.7.3 Specify Analog Ports

The command is used to identify top-level analog ports in a design.

set_analog_ports {ANA1 BANA}

9.7.4 Specify Diode Ports

This command is used to list the ports of a macro cell that connect to the positive or negative pins of a diode within a macro cell.

```
set_diode_ports \
  -positive {IN1 IN2} \
  -negative Z
```

9.7.5 Specify Floating Ports

The command specifies a list of ports of a macro cell that are not connected to any logic inside the macro cell.

set_floating_ports {portA portB portC}

9.7.6 Specify Pad Ports

The command specifies a list of pad ports of a macro cell.

set_pad_ports {PACK PSYNC}

9.7.7 Specify Power Source Reference Pin

The command specifies an input pin of the macro cell that is the voltage reference for a power domain.

```
set_power_source_reference_pin VREF \
  -domain PDCOP \
  -voltage_range 0.8:1.2
```

9.7.8   Specify Wire Feedthrough Ports

The command specifies a list of input and output ports that are internally connected by only a wire.

```
set_wire_feedthrough_ports {A B C}
# Ports A, B and C are internally connected
# by only a wire.
```

9.8   Version and Verification Support Commands

9.8.1   Specify CPF Version

The command specifies the version of CPF being used. This command if specified must be the very first command in the file.

```
set_cpf_version 2.0
# Default is 1.1.
```

9.8.2   Create Assertion Control

Assertions in a shutdown domain can be controlled using this command. The assertions can either be kept active (the default) or be turned off.

```
create_assertion_control \
  -name AC_1 \
  -assertions {A1 A2} \
  -type suspend \
  -shutoff_condition EN
```

The assertion control name is *AC_1*. It control two assertions *A1* and *A2*. These get suspended when the shutdown condition is true. If the shutdown condition is not specified, then the assertions suspend when the power domain in which *A1* and *A2* appear is shut down.

9.8.3   Specify Illegal Domain Configuration

The command asserts that the specified domain conditions and power mode conditions are illegal.

```
assert_illegal_domain_configurations \
   -name name \
   -domain_conditions \
      domain_name@nominal_condition_name \
   -group_modes group_name@power_mode_name
```

The **-domain_conditions** option specifies the nominal condition of each power domain that is not legal. The domain name must have been created earlier using the **create_power_domain** command. The nominal condition must have been created using **create_nominal_condition** command. The **-group_modes** option specifies the mode of each power mode control group. The group name must have been created using the **set_power_mode_control_group** command. The power mode must have been created using the **create_power_mode** command.

9.8.4   Specify Simulation Control

This command specifies the action to be taken during simulation when the power is turned off or restored.

```
set_sim_control \
   -domains {PD_CPU PD_ALU} \
   -action disable_isolation
```

The value **disable_isolation** with the -**action** option specifies that the isolation logic inferred from the CPF should be disregarded. Other possible values for this option are **power_up_replay**, **disable_corruption** and **disable_retention**.

9.9   CPF File Format

Here is a typical format of a CPF file.

```
set_cpf_version 1.1
set_design MYCHIP
set_hierarchy_separator /

# Define library sets:
define_library_set . . .
```

```
# Define always-on cells:
define_always_on_cell . . .

# Define isolation cells:
define_isolation_cell . . .

# Define power switch cells:
define_power_switch_cell . . .

# Create power nets and ground nets:
create_power_nets . . .
create_ground_nets . . .

# Create power domains:
create_power_domain . . .
update_power_domain . . .
create_global_connection . . .

# Define nominal operating conditions:
create_nominal_condition . . .
update_nominal_condition . . .

# Define power modes:
create_power_mode . . .

# Define rules for isolation logic insertion:
create_isolation_rule . . .
update_isolation_rules . . .

# Define rules for state retention registers:
create_state_retention_rule . . .
update_state_retention_rules . . .

# Define rules for power switch insertion:
create_power_switch_rule . . .
update_power_switch_rule . . .

# Define operating corners:
create_operating_corner . . .

end_design
```

9.10   Mychip Example

Here is the CPF for the *MYCHIP* example described in Chap. 8.

```
set_cpf_version 2.0
set_design MYCHIP
set_hierarchy_separator /
```

```
# Define library sets:
set LIB_PATH /home/bond/generic/ip
set LIB_LIST [list ${LIB_PATH}/stdcell_wcl.lib \
  ${LIB_PATH}/io_wcl.lib]
define_library_set -name WCL_LIBS \
  -libraries $LIB_LIST

# Define isolation cells:
define_isolation_cell \
  -cells {O2ISO_T50 A2ISO_T50} \
  -power VDD \
  -ground GND \
  -enable ISO \
  -valid_location to

# Define power switch cells:
define_power_switch_cell \
  -cells {HEADBUF_T50} \
  -type header \
  -power VDDG \
  -power_switchable VDD \
  -stage_1_enable SLEEP_IN \
  -stage_1_output SLEEP_OUT

# Create power nets and ground nets:
create_power_nets \
  -nets VDD1p0 \
  -voltage 1.0

create_power_nets \
  -nets VDDdsp \
  -voltage {0.9:1.1}

create_power_nets \
  -nets VDD0p9 \
  -voltage 0.9

create_ground_nets \
  -nets GND

# Create power domains:
create_power_domain \
  -name PD_MYCHIP \
  -default

update_power_domain \
  -name PD_MYCHIP \
  -primary_power_net VDD1p0 \
  -primary_ground_net GND
```

```
create_global_connection \
  -domain PD_MYCHIP \
  -net VDD1p0 \
  -pins PVDD1p0

create_power_domain \
  -name PD_CPU \
  -instances {U_CPU}

update_power_domain \
  -name PD_CPU \
  -primary_power_net VDD0p9 \
  -primary_ground_net GND

create_global_connection \
  -domain PD_CPU \
  -net VDD0p9 \
  -pins PVDD0p9

create_power_domain \
  -name PD_DSP \
  -instances {U_DSP}

update_power_domain \
  -name PD_DSP \
  -primary_power_net VDDdsp \
  -primary_ground_net GND

create_global_connection \
  -domain PD_DSP \
  -net VDDdsp \
  -pins PVDDdsp

create_power_domain \
  -name PD_COP \
  -instances {U_COP}

update_power_domain \
  -name PD_COP \
  -primary_power_net VDD1p0_SW \
  -primary_ground_net GND

create_global_connection \
  -domain PD_COP \
  -net VDD1p0 \
  -pins PVDD1p0

# Define nominal operating conditions:
create_nominal_condition \
  -name HIGHV \
```

```
  -voltage 1.1 \
  -state on

update_nominal_condition \
  -name HIGHV \
  -library_set BC_LIBS

create_nominal_condition \
  -name MEDV \
  -voltage 1.0 \
  -state on

update_nominal_condition \
  -name MEDV \
  -library_set TYP_LIBS

create_nominal_condition \
  -name LOWV \
  -voltage 0.9
  -state on

update_nominal_condition \
  -name LOWV \
  -library_set WCL_LIBS

create_nominal_condition \
  -name OFFV \
  -voltage 0 \
  -state off

# Define power modes:
create_power_mode \
  -name PM1 \
  -domain_conditions {PD_MYCHIP@MEDV PD_CPU@LOWV \
    PD_COP@MEDV PD_DSP@HIGHV}
create_power_mode \
  -name PM2 \
  -domain_conditions {PD_MYCHIP@MEDV PD_CPU@LOWV \
    PD_COP@OFFV PD_DSP@HIGHV}
create_power_mode \
  -name PM3 \
  -domain_conditions {PD_MYCHIP@MEDV PD_CPU@LOWV \
    PD_COP@MEDV PD_DSP@LOWV}
create_power_mode \
  -name PM4\
  -domain_conditions {PD_MYCHIP@MEDV PD_CPU@LOWV \
    PD_COP@OFFV PD_DSP@LOWV}
```

```
# Define rules for isolation logic insertion:
create_isolation_rule \
  -name PD_COP_ISO \
  -from PD_COP \
  -isolation_condition {U_PC/ISE} \
  -isolation_target from \
  -isolation_output low

update_isolation_rules \
  -name PD_COP_ISO \
  -location to \
  -prefix COP_ISO

# Define rules for state retention registers:
create_state_retention_rule \
  -name PD_COP_RET \
  -domain PD_COP \
  -restore_edge {U_PC/SRE}

update_state_retention_rules \
  -name PD_COP_RET \
  -cells {U_COP/reg1 U_COP/pc U_COP/int_state}

# Define rules for power switch insertion:
create_power_switch_rule \
  -name PD_COP_SW \
  -domain PD_COP \
  -external_power_net Vdd1p0

update_power_switch_rule \
  -name PD_COP_SW \
  -cells HEADBUF_T50 \
  -prefix COP_SW

# Define operating corners:
create_operating_corner \
  -name MAX_WCL \
  -voltage 0.9 \
  -library_set WCL_LIBS \
  -temperature 0 \
  -process 1

end_design
```

Appendix A. SAIF Syntax

This appendix describes the SAIF syntax for backward SAIF, library forward SAIF and RTL forward SAIF.

Complete Backward SAIF Syntax

Here is the complete syntax for a backward SAIF file. The starting nonterminal is *backward_saif_file*.

```
backward_leakage_spec ::=
  (LEAKAGE state_dep_timing_attributes
    {state_dep_timing_attributes} )

backward_instance_info ::=
  (INSTANCE [string] path {backward_instance_spec}
    {backward_instance_info} )
| (VIRTUAL_INSTANCE string path backward_port_spec )

backward_instance_spec ::=
  backward_net_spec
| backward_port_spec
| backward_leakage_spec

backward_net_info ::=
  (net_name net_switching_attributes)

backward_net_spec ::=
  (NET backward_net_info {backward_net_info})

backward_port_info ::=
  (port_name port_switching_attributes)
```

R. Chadha and J. Bhasker, *An ASIC Low Power Primer: Analysis,*
Techniques and Specification, DOI 10.1007/978-1-4614-4271-4,
© Springer Science+Business Media New York 2013

```
backward_port_spec ::=
  (PORT backward_port_info {backward_port_info})
```

```
backward_saif_file ::=
  (SAIFFILE backward_saif_header
    backward_saif_info)
```

```
backward_saif_header ::=
  backward_saif_version
  direction
  design_name
  date
  vendor
  program_name
  program_version
  hierarchy_divider
  time_scale
  duration
```

```
backward_saif_info ::= {backward_instance_info}
```

```
backward_saif_version := (SAIFVERSION string)
```

```
binary_operator ::= * | ^ | |
```

```
cond_expr ::=
  port_name
| unary_operator cond_expr
| cond_expr binary_operator cond_expr
| (cond_expr)
```

```
date ::= (DATE [string])
```

```
design_name ::= (DESIGN [string])
```

```
direction ::= (DIRECTION string)
```

```
duration ::= (DURATION rnumber)
```

```
edge_type := RISE | FALL
```

```
hierarchy_divider ::= (DIVIDER [hchar])
```

```
net_name ::= identifier
```

```
net_switching_attributes ::=
  {net_switching_attribute}
```

```
net_switching_attribute ::=
  simple_timing_attribute
| simple_toggle_attribute
```

```
path_dep_toggle_attributes ::=
  (path_dep_toggle_item {path_dep_toggle_item}
    [IOPATH_DEFAULT simple_toggle_attribute])

path_dep_toggle_item ::=
  IOPATH port_name {port_name}simple_toggle_attribute
port_name ::= identifier

port_switching_attributes ::=
  {port_switching_attribute}

port_switching_attribute ::=
  simple_timing_attribute
| simple_toggle_attribute
| state_dep_toggle_attributes
| path_dep_toggle_attributes
| sdpd_toggle_attributes

potential_pd_toggle_attributes :=
  path_dep_toggle_attributes
| simple_toggle_attribute

program_name ::= (PROGRAM_NAME [string])

program_version ::= (PROGRAM_VERSION [string])

timeunit ::= s | ms | us | ns | ps | fs

time_scale ::= (TIMESCALE [dnumber timeunit])

sdpd_default_toggle_item ::=
  COND_DEFAULT potential_pd_toggle_attributes
| COND_DEFAULT (edge_type)
    potential_pd_toggle_attributes
    [COND_DEFAULT (edge_type)
    potential_pd_toggle_attributes]

sdpd_toggle_attributes ::=
  (sdpd_toggle_item {sdpd_toggle_item}
    [sdpd_default_toggle_item])

sdpd_toggle_item ::=
  COND cond_expr [(edge_type)]
    potential_pd_toggle_attributes

sd_simple_timing_attributes ::=
  {sd_simple_timing_attribute}

sd_simple_timing_attribute ::=
  (T1 rnumber)
| (T0 rnumber)
```

```
simple_timing_attribute ::=
  (T0 rnumber)
| (T1 rnumber)
| (TX rnumber)
| (TZ rnumber)
| (TB rnumber)

simple_toggle_attribute ::=
  (TC rnumber)
| (TG rnumber)
| (IG rnumber)
| (IK rnumber)

state_dep_default_toggle_item ::=
  COND_DEFAULT simple_toggle_attribute
| COND_DEFAULT (edge_type) simple_toggle_attribute
    [COND_DEFAULT (edge_type)
      simple_toggle_attribute]

state_dep_timing_attributes ::=
  (state_dep_timing_item {state_dep_timing_item}
    [COND_DEFAULT sd_simple_timing_attributes])

state_dep_timing_item ::=
  COND cond_expr sd_simple_timing_attributes

state_dep_toggle_attributes ::=
  (state_dep_toggle_item {state_dep_toggle_item}
    [state_dep_default_toggle_item])

state_dep_toggle_item :=
  COND cond_expr [(edge_type)]
    simple_toggle_attribute

unary_operator ::= !

vendor ::= (VENDOR [string])
```

Complete RTL Forward SAIF Syntax

Here is the complete syntax for the RTL forward SAIF file. The starting nonterminal is *rforward_saif_file*.

```
date ::=
  (DATE [string])
```

```
design_name ::=
  (DESIGN [string])

direction ::=
  (DIRECTION string)

hierarchy_divider ::=
  (DIVIDER [hchar])

instance_name ::=
  hierarchical_identifier

mapped_name ::=
  hierarchical_identifier

net_mapping_directives ::=
  (NET {(rtl_name mapped_name)})

port_mapping_directives ::=
  (PORT {(rtl_name mapped_name [string])})

program_name ::=
  (PROGRAM_NAME [string])

program_version ::=
  (PROGRAM_VERSION [string])

rforward_saif_file ::=
  (SAIFILE rforward_saif_header rforward_saif_info)

rforward_saif_header ::=
  rforward_saif_version
  direction
  design_name
  date
  vendor
  program_name
  program_version
  hierarchy_divider

rforward_saif_info ::=
  {rforward_instance_declaration}

rforward_instance_declaration ::=
  (INSTANCE [string] instance_name
    {rforward_instance_directive}
    {rforward_instance_declaration})

rforward_instance_directive ::=
  port_mapping_directives
| net_mapping_directives
```

```
rforward_saif_version ::=
  (SAIFVERSION string)

rtl_name ::=
  hierarchical_identifier

vendor ::=
  (VENDOR [string])
```

Complete Library Forward SAIF Syntax

Here is the complete syntax for the library forward SAIF file. The starting nonterminal is *lforward_saif_file*.

```
date ::=
  (DATE [string])

design_name ::=
  (DESIGN [string])

direction ::=
  (DIRECTION string)

hierarchy_divider ::=
  (DIVIDER [hchar])

leakage_declaration ::=
  (LEAKAGE {state_dep_timing_directive})

lforward_saif_file ::=
  (SAIFILE lforward_saif_header lforward_saif_info)

lforward_saif_header ::=
  lforward_saif_version
  direction
  design_name
  date
  vendor
  program_name
  program_version
  hierarchy_divider

lforward_saif_version ::=
  (SAIFVERSION string [string])

library_sdpd_info ::=
  (LIBRARY string [string]
    {module_sdpd_declaration})
```

```
module_name ::=
  identifier

module_sdpd_declaration ::=
  (MODULE module_name {module_sdpd_directive})

module_sdpd_directive ::=
  port_declaration
| leakage_declaration

path_dep_toggle_directive ::=
  (path_dep_toggle_directive_item
    {path_dep_toggle_directive_item}
    [IOPATH_DEFAULT])

path_dep_toggle_directive_item ::=
  IOPATH port_name {port_name}

port_declaration ::=
  (PORT port_name {port_directive})

port_directive ::=
  state_dep_toggle_directive
| path_dep_toggle_directive
| sdpd_toggle_directive

program_name ::=
  (PROGRAM_NAME [string])

program_version ::=
  (PROGRAM_VERSION [string])

sdpd_toggle_directive ::=
  (sdpd_toggle_directive_item
    {sdpd_toggle_directive_item}
    [COND_DEFAULT [RISE_FALL]]
    [path_dep_toggle_directive]])

sdpd_toggle_directive_item ::=
  COND cond_expr [RISE_FALL]
    [path_dep_toggle_directive]

state_dep_timing_directive ::=
  (state_dep_timing_directive_item
    {state_dep_timing_directive_item}
    [COND_DEFAULT])

state_dep_timing_directive_item ::=
  COND cond_expr
```

```
state_dep_toggle_directive ::=
  (state_dep_toggle_directive_item
    {state_dep_toggle_directive_item}
    [COND_DEFAULT [RISE_FALL]])

state_dep_toggle_directive_item ::=
  COND cond_expr [RISE_FALL]

vendor ::=
  (VENDOR [string])
```

Appendix B. UPF Syntax

This appendix provides a complete alphabetical description of all the commands in UPF.[1] Keywords are in **bold** font. Values that need to be provided by user are in *italic* font. Square brackets ([ ]) show optional parameters. Bold square brackets (**[ ]**) and bold curly braces (**{ }**) are required characters. Curly braces ({ }) indicate a list of parameters. An asterisk (*) implies parameters that can be repeated. Angle brackets (< >) indicate grouping of alternate values. Finally the separator bar (|) indicates optional choices.

```
# Adds design elements to a power domain:
add_domain_elements domain_name
  -elements element_list

# Adds the states to the port:
add_port_state port_name
  {-state
    {name <nom | min max | min nom max | off>}}*

# Attributes power state to a power domain
# or a supply net:
add_power_state object_name
  {-state state_name
    {[-supply_expr {boolean_function}]
     [-logic_expr {boolean_function}]
     [-simstate simstate]
     [-legal | -illegal] [-update]}}*
  [-simstate simstate]
  [-legal | -illegal]
  [-update]
```

---

[1] Note that not all tools support all the options as of this book's publication.

R. Chadha and J. Bhasker, *An ASIC Low Power Primer: Analysis, Techniques and Specification*, DOI 10.1007/978-1-4614-4271-4, © Springer Science+Business Media New York 2013

```
# Defines the states of each of the supply nets:
add_pst_state state_name
  -pst table_name
  -state supply_states

# Associates a supply set to a power domain:
associate_supply_set supply_set_ref
  -handle supply_set_handle

# Inserts checker modules and binds them to
# design elements:
bind_checker instance_name
  -module checker_name
  [-elements element_list]
  [-bind_to module [-arch name]]
  [-ports {{port_name net_name}*}]

# Connects a logic net to logic ports:
connect_logic_net net_name
  -ports port_list

# Connects a supply net to supply ports:
connect_supply_net net_name
  [-ports list]
  [-pg_type {pg_type_list element_list}]*
  [-vct vct_name]
  [-pins list]
  [-cells list]
  [-domain domain_name]
  [-rail_connection rail_type]

# Connects a supply set to specified elements:
connect_supply_set supply_set_ref
  {-connect {supply_function {pg_type_list}}}*
  [-elements elements_list]
  [-exclude_elements exclude_list]
  [-transitive <TRUE | FALSE>]

# Defines a composite domain comprised of
# one or more subdomains:
create_composite_domain composite_domain_name
  [-subdomains subdomain_list]
  [-supply {supply_set_handle [supply_set_ref]}]*
  [-update]
```

```
# Defines a value conversion table that can be used
# to convert HDL logic values into
# state type values:
create_hdl2upd_vct vct_name
  -hdl_type {<vhdl | sv > [typename]}
  -table {{from_value to_value}*}

# Defines a logic net:
create_logic_net net_name

# Defines a logic port:
create_logic_port port_name
  [-direction <in | out | inout>]

# Defines the set of elements in a power domain:
create_power_domain domain_name
  [-simulation_only]
  [-elements element_list]
  [-exclude_elements exclude_list]
  [-include scope]
  [-supply {supply_set_handle [supply_set_ref]}*]]
        [-scope instance_name]
  [-define_func_type {supply_function
    {pg_type_list}}]*
  [-update]

# Define a power switch:
create_power_switch switch_name
  -output_supply_port {port_name
    [supply_net_name]}
  {-input_supply_port {port_name
    [supply_net_name]}}*
  {-control_port {port_name [net_name]}}*
  {-on_state {state_name input_supply_port
    {boolean_function}}}*
  [-off_state {state_name {boolean_function}}]*
  [-supply_set supply_set_name]
  [-on_partial_state {state_name input_supply_port
    {boolean_function}}]*
  [-ack_port {port_name net_name
    [{boolean_function}]}]*
  [-ack_delay {port_name_delay}]*
  [-error_state {state_name {boolean_function}}]*
  [-domain domain_name]

# Creates a power state table:
create_pst table_name
  -supplies supply_list
```

```
# Creates a supply net:
create_supply_net net_name
  [-domain domain_name]
  [-reuse]
  [-resolve <unresolved | one_hot | parallel |
    parallel_one_hot>]

# Creates a supply port on a design element:
create_supply_port port_name
  [-domain domain_name]
  [-direction <in | out | inout>]

# Creates a supply set:
create_supply_set set_name
  [-function {func_name [net_name]}]*
  [-reference_gnd supply_net_name]
  [-update]

# Defines value conversion table to be used in
# converting UPF values into HDL logic values:
create_upf2hdl_vct vct_name
  -hdl_type {<vhdl | sv> [typename]}
  -table {{from_value to_value}*}

# Describes a legal state transition:
describe_state_transition transition_name
  -object object_name
  {-from {from_list} -to {to_list} |
    -paired {{from_state to_state}*} |
    -from {from_list} -to {to_list}
      -paired {{from_state to_state}*}}
  [-legal | -illegal]

# Loads the simstate behavior from
# set_simstate_behavior commands:
load_simstate_behavior lib_name
  -file {file}*

# Loads in the UPF file at the scope specified:
load_upf upf_file_name
  [-scope instance_name]
  [-version upf_version]

# Loads the UPF file without modifying
# the global variables:
load_upf_protected upf_file_name
  [-hide_globals]
  [-scope scope_name]
```

```
    [-version upf_version]
    [-params param_list]

  # Specifies which library cell to use
  # for isolation cell:
  map_isolation_cell isolation_name
    -domain domain_name
    [-elements element_list]
    [-lib_cells lib_cells_list]
    [-lib_cell_type lib_cell_type]
    [-lib_model_name model_name
      {-port {port_name net_name}}*]

  # Specifies which library cell to use
  # for level shifter:
  map_level_shifter_cell level_shifter_strategy
    -domain domain_name
    -lib_cells list
    [-elements element_list]

  # Specifies which library cell to use
  # for power switch:
  map_power_switch {switch_name}
    -domain domain_name
    -lib_cells {list}
    [-port_map {{mapped_model_port
        switch_port_or_supply_net_ref}*}]

  # Specifies which library cell to use
  # for retention cell:
  map_retention_cell retention_name_list
    -domain domain_name
    [-element element_list]
    [-exclude_elements exclude_list]
    [-lib_cells lib_cell_list]
    [-lib_cell_type lib_cell_type]
    [-lib_model_name name
      {-port port_name net_ref}*]

  # Merge two or more power domains into a
  # new power domain:
  merge_power_domains new_domain_name
    -power_domains list
    [-scope instance_name]
    [-all_equivalent]
```

```
# Defines the format to be used for new names that
# are implicitly created:
name_format
  [-isolation_prefix string]
  [-isolation_suffix string]
  [-level_shift_prefix string]
  [-level_shift_suffix string]
  [-implicit_supply_suffix string]
  [-implicit_logic_prefix string]
  [-implicit_logic_suffix string]

# Creates a UPF file with the current commands
# active in scope:
save_upf upf_file_name
  [-scope instance_name]
  [-version string]

# Applies attributes to design elements:
set_design_attributes
  < -elements element_list |
    -models model_list |
    -elements element_list -models model_list
      -exclude_elements exclude_list |
    -exclude_elements exclude_list
      -models model_list >
  [-attribute name value]*

# Specifies the design root:
set_design_top root

# Sets the default power and ground supply nets
# for a power domain:
set_domain_supply_net domain_name
  -primary_power_net supply_net_name
  -primary_ground_net supply_net_name

# Defines an isolation strategy:
set_isolation isolation_name
  -domain ref_domain_name
  [-elements element_list]
  [-source source_supply_ref
    | -sink sink_supply_ref
    | -source source_supply_ref
      -sink sink_supply_ref
    | -applies_to <inputs | outputs | both>]
  [-applies_to_clamp <0 | 1 | any | Z | latch |
      value>]
```

```
    [-applies_to_sink_off_clamp <0 | 1 | any |
       Z |latch | value>]
    [-applies_to_source_off_clamp <0 | 1 | any |
       Z | latch | value>]
    [-isolation_power_net net_name]
    [-isolation_ground_net net_name]
    [-no_isolation]
    [-isolation_supply_set supply_set_list]
    [-isolation_signal signal_list
      [-isolation_sense {<high | low>*}]]
    [-name_prefix string]
    [-name_suffix string]
    [-clamp_value {<0 | 1| any | Z | latch | value>*}]
    [-sink_off_clamp <0 | 1 | any | Z | latch |
       value> [simstate_list]]
    [-source_off_clamp <0 | 1 | any | Z | latch |
       value> [simstate_list]]
    [-location <automatic | self | fanout | fanin |
       faninout | parent | sibling>]
    [-force_isolation]
    [-instance {{instance_name port_name}*}]
    [-diff_supply_only <TRUE | FALSE>]
    [-transitive <TRUE | FALSE>]
    [-update]

# Specifies the control signal for an
# isolation strategy:
set_isolation_control isolation_name
  -domain domain_name
  -isolation_signal signal_name
  [-isolation_sense <high | low>]
  [-location <self | parent | sibling | fanout |
    automatic>]

# Specifies a level shifter strategy:
set_level_shifter level_shifter_name
  -domain domain_name
  [-elements element_list]
  [-no_shift]
  [-threshold value | list]
  [-force_shift]
  [-source domain_name]
  [-sink domain_name]
  [-applies_to <inputs | outputs | both>]
  [-rule <low_to_high | high_to_low | both>]
  [-location <self | parent | sibling | fanout |
    automatic>]
```

```
[-name_prefix string]
[-name_suffix string]
[-input_supply_set supply_set_name]
[-output_supply_set supply_set_name]
[-internal_supply_set supply_set_name]
[-instance {{instance_name port_name}*}]
[-transitive <TRUE | FALSE>]
[-update]
```

\# Defines how PARTIAL_ON is to be interpreted
\# in specified tools:
```
set_partial_on_translation [OFF | FULL_ON]
[-full_on_tools {string_list}]
[-off_tools {string_list}]
```

\# Defines the power and ground pins
\# of a library cell:
```
set_pin_related_supply library_cell
 -pins list
 -related_power_in supply_pin
 -related_ground_pin supply_pin
```

\# Defines information on ports:
```
set_port_attributes
  [-ports {port_list}]
  [-exclude_ports {ports_list}]
  [{-domains {domain_list} [-applies_to <inputs |
     outputs | both>]}]
  [{-exclude_domains {domain_list} [-applies_to
     <inputs | outputs | both>]}]
  [{-elements {element_list} [-applies_to <inputs
     | outputs | both>]}]
  [{-exclude_elements {exclude_list} [-applies_to
     <inputs | outputs | both>]}]
  [-model name]
  [-attribute name_value]*
  [-clamp_value <0 | 1 | any | Z | latch | value>]
  [-sink_off_clamp <0 | 1 | any | Z | latch| value>]
  [-source_off_clamp <0 | 1 | any | Z | latch
    | value>]
  [-receiver_supply supply_set_ref]
  [-driver_supply supply_set_ref]
  [-related_power_port supply_port]
  [-related_ground_port supply_port]
  [-related_bias_ports supply_port_list]
  [-repeater_supply supply_set_ref]
```

```
    [-pg_type pg_type_value]
    [-transitive <TRUE | FALSE>]

 # Adds additional information to a power switch:
 set_power_switch switch_name
    -output_supply_port
       {port_name [supply_net_name]}
    {-input_supply_port
       {port_name [supply_net_name]}}*
    {-control_port {port_name}}*
    {-on_state {state_name input_supply_port
       {boolean_function}}}*
    [-supply_set supply_set_name]
    [-on_partial_state {state_name input_supply_port
       {boolean_function}}]*
    [-off_state {state_name {boolean_function}}]*
    [-error_state {state_name {boolean_function}}]*

 # Specifies retention strategy:
 set_retention retention_name
    -domain domain_name
    [-elements element_list]
    [-exclude_elements exclude_list]
    [-retention_power_net net_name]
    [-retention_ground_net net_name]
    [-retention_supply_net ret_supply_net]
    [-no_retention]
    [-save_signal {{logic_net <high | low |
       posedge | negedge>}}
    -restore_signal {{logic_net <high | low |
       posedge | negedge>}}]
    [-save_condition {{boolean_function}}]
    [-restore_condition {{boolean_function}}]
    [-use_retention_as_primary]
    [-parameters {< <RET_SUP_COR | NO_RET_SUP_COR> |
       <SAV_RES_COR | NO_SAVE_RES_COR> >*}]
    [-instance {{instance_name [signal_name]}*}]
    [-transitive <TRUE | FALSE>]
    [-update]

 # Specifies the control signals for a
 # retention strategy:
 set_retention_control retention_name
    -domain domain_name
    -save_signal {{net_name <high | low |
       posedge | negedge>}}
```

```
  -restore_signal {{net_name <high| low |
     posedge | negedge>}}
  [-assert_r_mutex {{net_name <high | low |
     posedge | negedge>}}]*
  [-assert_s_mutex {{net_name <high | low |
     posedge | negedge>}}]*
  [-assert_rs_mutex {{net_name <high | low |
     posedge | negedge>}}]*

# Creates a named list of retention elements:
set_retention_elements retention_list_name
  [-elements element_list]
  [{-applies_to <required | not_optional |
     not_required | optional>}]
  [-exclude_elements exclude_list]
  [-retention_purpose <required | optional>]
  [-transitive <TRUE | FALSE>]

# Specifies the scope for which the UPF
# commands apply:
set_scope instance

# Specifies the simulation simstate behavior
# for cells in a library:
set_simstate_behavior <ENABLE | DISABLE>
  [-lib name]
  [-model list]

# Specifies the UPF version for UPF commands:
upf_version [string]

# Specifies the functional model for isolation
# and level shifting:
use_interface_cell interface_implementation_name
  -strategy
     list_of_isolation_level_shifter_strategies
  -domain domain_name
  -lib_cells lib_cell_list
  [-map {{port net_ref}*}]
  [-elements element_list]
  [-exclude_elements exclude_list]
  [-applies_to_clamp
     <0 | 1 | any | Z | latch | value>]
  [-update_any <0 | 1 | known | Z | latch | value>]
  [-force_function]
  [-inverter_supply_set list]
```

References

[BUC] Buch K., *HDL Design Methods for Low-Power Implementation*, ESNUG.

[BHA06] Bhasker J., *The Exchange Format Handbook: A DEF, LEF, PDEF, SDF, SPEF & VCD Primer*, Star Galaxy Publishing, 2006, ISBN 0-9650391-3-7.

[BHA10] Bhasker J., *A SystemVerilog Primer*, Star Galaxy Publishing, 2010, ISBN 978-0-9650391-1-6.

[CPF11] S*i2 Common Power Format Specification*, Version 2.0, Feb. 2011, Si2 Inc., www.si2.org.

[ESM11] Esmaeilzadeh H., et. al., *Dark Silicon and the End of Multicore Scaling*, ISCA'11, ACM, 2011.

[FLY07] Flynn D., et. al., *Low Power Methodology Manual: For System-on-Chip Design*, Springer, 2007.

[JED10] *DDR3 SDRAM Specification*, JESD79-3E, July 2010.

[LIB] *Liberty Users Guide*, available at "http://www.opensourceliberty.org".

[MOO65] Moore G., *Cramming More Components onto Integrated Circuits*, Electronics, Vol. 38, No. 8, 1965.

[MUK86] Mukherjee A., *Introduction to nMOS and CMOS VLSI Systems Design*, Prentice Hall, 1986.

[RAB09] Rabaey J., *Low Power Design Essentials*, Springer, 2009.

[STR05] Streetman B. and S. Bannerjee, *Solid-state Electronic Devices, 6th edition*, Prentice Hall, 2005.

[SUB10] Subramaniam P., *Power Management for Optimal Power Design*, EDN, May 2010.

[SZE81] Sze S.M., *Physics of Semiconductor Devices, Second Edition*, John Wiley & Sons, 1981.

[UPF09] *IEEE Std for Design and Verification of Low Power Integrated Circuits*, IEEE Std 1801-2009, IEEE.

R. Chadha and J. Bhasker, *An ASIC Low Power Primer: Analysis, Techniques and Specification*, DOI 10.1007/978-1-4614-4271-4,
© Springer Science+Business Media New York 2013

Index

R. Chadha and J. Bhasker, *An ASIC Low Power Primer: Analysis, Techniques and Specification*, DOI 10.1007/978-1-4614-4271-4,
© Springer Science+Business Media New York 2013